PRINCIPLES AND LABS FOR DEEP LEARNING

PRINCIPLES AND LABS FOR DEEP LEARNING

SHIH-CHIA HUANG

Professor, Department of Electronic Engineering, National Taipei University of Technology, Taipei, Taiwan

TRUNG-HIEU LE

Assistant Professor, Department of Electronic Engineering, National Taipei University of Technology, Taipei, Taiwan
Lecturer, Faculty of Information Technology, Hung Yen University of Technology and Education, Hung Yen, Vietnam

ACADEMIC PRESS

An imprint of Elsevier

ELSEVIER

Academic Press is an imprint of Elsevier
125 London Wall, London EC2Y 5AS, United Kingdom
525 B Street, Suite 1650, San Diego, CA 92101, United States
50 Hampshire Street, 5th Floor, Cambridge, MA 02139, United States
The Boulevard, Langford Lane, Kidlington, Oxford OX5 1GB, United Kingdom

Library of Congress Cataloging-in-Publication Data
A catalog record for this book is available from the Library of Congress

British Library Cataloguing-in-Publication Data
A catalogue record for this book is available from the British Library

ISBN 978-0-323-90198-7

For information on all Academic Press publications
visit our website at https://www.elsevier.com/books-and-journals

Publisher: Mara Conner
Acquisitions Editor: Chris Katsaropoulos
Editorial Project Manager: Isabella C. Silva
Production Project Manager: Swapna Srinivasan
Cover Designer: Greg Harris

Typeset by SPi Global, India

Contents

Preface

In 1943, Warren McCulloch and Walter Pitts introduced a computational model based on a threshold logic algorithm, paving the way for the development of artificial intelligence (AI). The field of AI research was founded at the Dartmouth summer research project on artificial intelligence in 1956. In the nearly 80 years of development history, AI has experienced many ups and downs, especially the two "AI winters," which are known as the periods of reduced funding and interest in AI research. After AlphaGo, a deep learning-based computer Go program developed by Google DeepMind, defeated a professional Go player in 2015, AI has once again attracted considerable research attention. In recent years, AI with deep learning algorithms has been dramatically improved and successfully applied to many problems in academia and industry. Recognizing the potential and power of deep learning, many world-class technology companies, such as Google, Facebook, Microsoft, Tesla, and others, have invested a lot of manpower and resources in researching and developing their products using deep learning.

The authors of this book deeply understand the importance and development of the field of AI. As such, *Principles and Labs for Deep Learning* is meant to inspire and help more students, engineers, and researchers to quickly enter the field of AI and begin applying deep learning in their research projects, products, and platforms. This book includes 12 chapters with content that balances theory and practice.

Chapter 1: "Introduction to TensorFlow 2"
Chapter 2: "Neural networks"
Chapter 3: "Binary classification problem"
Chapter 4: "Multi-category classification problem"
Chapter 5: "Training neural network"
Chapter 6: "Advanced TensorFlow"
Chapter 7: "Advanced TensorBoard"
Chapter 8: "Convolutional neural network architectures"
Chapter 9: "Transfer learning"
Chapter 10: "Variational auto-encoder"
Chapter 11: "Generative adversarial network"
Chapter 12: "Object detection"

The first half of each chapter introduces and analyzes the corresponding theories to help readers understand the core fundamentals of deep learning. The last half of each chapter carefully designs and implements the example programs step by step to help readers reinforce what they just learned. The most popular open-source library, TensorFlow, is employed to implement example programs. Lastly, each chapter has one corresponding Lab as well as instructions for implementation to guide readers in practicing and accomplishing specific learning outcomes. All Labs in this book can be downloaded from GitHub: https://github.com/taipeitechmmslab/MMSLAB-DL/tree/master.

Shih-Chia Huang and Trung-Hieu Le
Department of Electronic Engineering,
National Taipei University of Technology, Taipei, Taiwan

Environment installation

OUTLINE
- Installation of Python, TensorFlow, and expansion packages on the Windows and Ubuntu operating systems
- Installation of Jupyter Notebook and PyCharm IDE on Windows and Ubuntu operating system
- Downloading all the example programs of this book from GitHub, and running the code through Jupyter Notebook

1 Python installation

1.1 Windows environment

To install Python, first go to the official Python website www.python.org/downloads/windows/, then download the Python installation file. The version of TensorFlow requires Python3.4, Python3.5, or Python3.6, so be sure to install any of these three versions. The following will use Python3.6.8 as an example.

1. Download the 3.6.8 installation file, as shown in Fig. 1.

- Python 3.6.8 - Dec. 24, 2018

Note that Python 3.6.8 *cannot* be used on Windows XP or earlier.

- Download Windows help file
- Download Windows x86-64 embeddable zip file
- Download Windows x86-64 executable installer
- Download Windows x86-64 web-based installer
- Download Windows x86 embeddable zip file
- Download Windows x86 executable installer
- Download Windows x86 web-based installer

FIG. 1   Download the Python installation file.

2. Install Python: Install Python with default settings, as shown in Fig. 2.

FIG. 2 Installation of Python.

1.2 Ubuntu environment

Since the Ubuntu operating system has built-in Python, the installation of Python in this section is omitted.

2 TensorFlow installation

There are many methods for installing TensorFlow, such as using local installation, Virtualenv virtual machine, Docker, and so on. This book uses the Virtualenv virtual machine recommended by TensorFlow, and introduces the installation of TensorFlow with CPU and GPU supports on two operating systems including Windows and Ubuntu.

2.1 Windows environment

1. Open the command prompt: Press "Windows logo key + R" to open "Run" box, then type "cmd" and click "OK" as shown in Fig. 3.

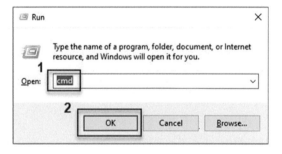

FIG. 3 Open command prompt.

2. Install the virtualenv virtual machine:

```
pip install virtualenv
```

3. Create a new virtual environment: In the command line below, tf2 represents the name of the virtual environment.

```
virtualenv --system-site-packages -p python ./tf2
```

4. Enter the virtual environment (tf2), as shown in Fig. 4.

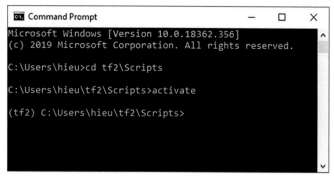

FIG. 4   Enter the virtual environment.

```
cd tf2\Scripts
activate
```

5. Upgrade pip version:

```
pip install --upgrade pip
```

6. Install TensorFlow: There are two versions of TensorFlow installation: TensorFlow CPU support and TensorFlow GPU support. If a computer has an Nvidia GPU and support CUDA, it is recommended to install the TensorFlow GPU support.

- TensorFlow CPU support:

```
pip install tensorflow
```

- TensorFlow GPU support:
  First, go to https://developer.nvidia.com/cuda-gpus to see whether the graphics card supports CUDA. Then install the graphics driver, CUDA10 and cuDNN, and TensorFlow GPU version in order.
  - **Install the graphics card driver**: Go to www.nvidia.com/download/index.aspx?lang=en-us to download the corresponding graphics card driver, as shown in Fig. 5, and install it.

FIG. 5   Download the graphics card driver.

- **CUDA**: Go to https://developer.nvidia.com/cuda-downloads, click the "Archive of Previous CUDA Releases" in the window, as shown in Fig. 6.

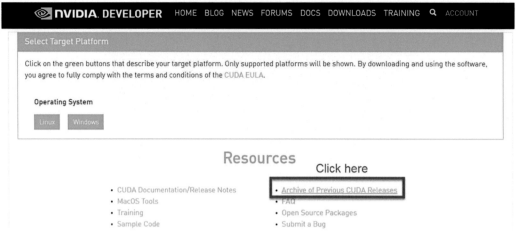

FIG. 6   CUDA download page.

- Select the version of CUDA Toolkit 10.0, as shown in Fig. 7.

FIG. 7   Select CUDA installation version.

- Select the installation package of Windows 10, download and install it, as shown in Fig. 8.

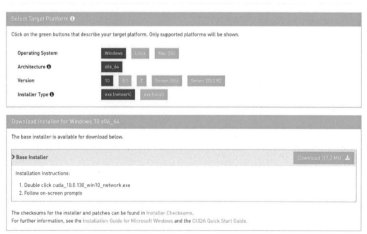

FIG. 8   Download the CUDA installation file.

- **cuDNN**: Go to https://developer.nvidia.com/rdp/cudnn-download to download cuDNN, as shown in Fig. 9. Note that, it is required to log in for downloading.

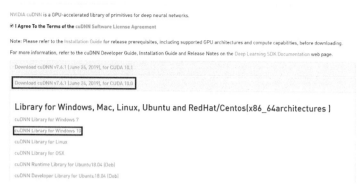

FIG. 9 Download the cuDNN installation file.

Unzip and copy three files from the unzipped folder of cuDNN to the CUDA installation directory. The steps are as follows:

- Copy the < cuDNN unzipped folder > \cuda\bin\cudnn64_7.dll file to "C:\ Program Files\NVIDIA GPU Computing Toolkit\CUDA\v10.0\bin" directory
- Copy the <cuDNN unzipped folder >\cuda\include\cudnn.h file to "C:\ Program Files\NVIDIA GPU Computing Toolkit\CUDA\v10.0\include" directory
- Copy the < cuDNN unzipped folder> \cuda\lib\x64\cudnn.lib file to "C:\ Program Files\NVIDIA GPU Computing Toolkit\CUDA\v10.0\lib\x64" directory

- **Install TensorFlow GPU support**:

```
pip install tensorflow-gpu
```

7. Verify the installation:

```
python -c "import tensorflow as tf;
print(tf.constant([[1, 2], [3, 4]]))"
```

2.2 Ubuntu environment

1. Open Terminal, as shown in Fig. 10.

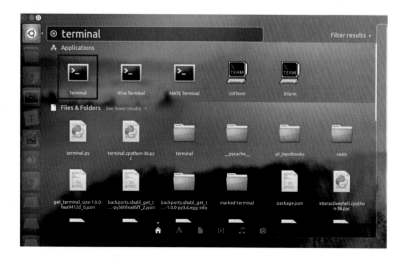

FIG. 10 Open terminal.

2. Install pip:

```
sudo apt-get install python3-pip
```

3. Install Virtualenv virtual machine:

```
sudo apt install virtualenv
```

4. Create a new virtual environment: In the command line below, tf2 represents the name of the virtual environment.

```
virtualenv --system-site-packages -p python3 ./tf2
```

5. Enter the virtual environment (tf2):

```
source tf2/bin/activate
```

6. Upgrade pip version:

```
pip install --upgrade pip
```

7. Install TensorFlow: There are two versions of TensorFlow installation: TensorFlow CPU support and TensorFlow GPU support.
   - TensorFlow CPU support:

```
pip install tensorflow
```

   - TensorFlow GPU support:
     - First, go to https://developer.nvidia.com/cuda-gpus to check whether the graphics card supports CUDA or not, then perform the following steps to install.
     - Second, go to https://www.tensorflow.org/install/gpu#ubuntu_ and follow the instruction to install drivers and libraries.

- Install TensorFlow GPU support using pip package:

```
pip install tensorflow-gpu
```

8. Verify the installation:

```
python -c "import tensorflow as tf;
print(tf.constant([[1, 2], [3, 4]]))"
```

Supplementary explanation

Compare with the environment where "local Python" is directly installed, multiple installation environments can be created through "Virtualenv virtual machine" or "Docker," in which different versions of TensorFlow can be installed in different environments without affecting each other. Thus, it is highly recommended to use the "Virtualenv virtual machine" or "Docker" for the installation of software packages.

3 Python extension installation

The installation of extension packages on Windows and Ubuntu is the same. First, open Terminal on Ubuntu or command prompt on Windows, then enter the Virtualenv virtual environment that installed, and finally, type the command in the window to install.

- Numpy: An open-source library for Python, which works with arrays and supports large and multi-dimensional matrices. To install Numpy, run the following:

```
pip install numpy
```

- Matplotlib: A plotting library for Python, which supports drawing both static and animated visualization. To install Matplotlib, run the following:

```
pip install matplotlib
```

- Pandas: A library built on top of Python for data manipulation and analysis. To install Pandas, run the following:

```
pip install pandas
```

- OpenCV: An open-source library, providing a common infrastructure for applications of computer vision such as image processing, object detection, facial recognition, and so on. To install OpenCV, run the following:

```
pip install opencv-python
```

- TensorFlow Addons: An expansion package of TensorFlow, which contains many new functions are not added to the core TensorFlow due to the limitation of their broad applicability or because they are not common. To install TensorFlow Addons, run the following:

```
pip install tensorflow-addons
```

- TensorFlow Datasets: A collection of datasets that are ready for use with a specific purpose such as developing, training, or testing machine learning models. All datasets can be obtained through tf.data.Datasets API. To install TensorFlow Datasets, run the following:

```
pip install tensorflow-datasets
```

- TensorFlow Hub: A repository of trained machine learning and deep learning models that can be reused by applying transfer learning or fine-tuning techniques to address new tasks with less training data and time. To install TensorFlow Hub, run the following.

```
pip install tensorflow-hub
```

4 Jupyter notebook

Jupyter Notebook is a web-based interactive application that supports creating and sharing documents containing live code, narrative text, and visualizations. The name "Jupyter" is derived from the three core programming languages including Julia, Python, and R. Through Jupyter Notebook, a lot of kernels that allow programming in more than forty languages can be connected. The IPython kernel, which allows writing source code in Python, is set as the default in Jupyter Notebook. The installation of the Jupyter Notebook on Windows and Ubuntu operating systems is introduced below.

4.1 Jupyter notebook installation

1. Windows environment
   - Install Jupyter Notebook: Please run the following.

   ```
   python -m pip install jupyter
   ```

   - Add the virtual environment to Jupyter Notebook: tf2 in the command line below represents the name of the virtual environment.

   ```
   .\tf2\Scripts\activate
   pip install ipykernel
   python -m ipykernel install --name=tf2
   ```

2. Ubuntu environment
   - Install Jupyter Notebook: please run the following.

   ```
   python3 -m pip install jupyter
   ```

   - Add the virtual environment to Jupyter Notebook: tf2 in the command line below represents the name of the virtual environment.

   ```
   source tf2/bin/activate
   pip3 install ipykernel
   python3 -m ipykernel install --user --name=tf2
   ```

4.2 Setup and create new project

1. Open Jupyter Notebook: please run the following.

   ```
   jupyter notebook
   ```

2. Create an execution file:
   Click the "New" icon on the top-right corner of the window and then click on the Python interpreter that installed, which is called kernel in Jupyter, to start, as shown in Fig. 11. There are three different kernels as follows:

   - Python3: Local Python
   - tf2: Virtual machine Python, which includes the TensorFlow-cpu version installed earlier
   - tf2-gpu: Virtual machine Python, which includes the TensorFlow-gpu version installed earlier

FIG. 11 Create an execution file.

3. Select kernel: After entering the Python file, choose Kernel → Change kernel → select kernel. The top-right corner will show which kernel is currently used, as shown in Fig. 12.

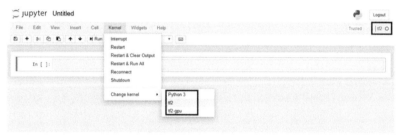

FIG. 12 Selection of kernel.

4.3 Jupyter Notebook operation

1. Notebooks modes: Jupyter Notebook has two modes including Edit Mode and Command Mode.
   - Edit mode: The mode of writing the code, which can be identified by the cell with a green border and green left margin, as shown in Fig. 13.

FIG. 13 Edit Mode.

To enter the edit mode, press the "Enter" key or directly click in a cell. After entering, to test this mode, type print ("Hello Jupyter Notebook") in the cell, then press "Shift + Enter" to run the code and see the result, as shown in Fig. 14.

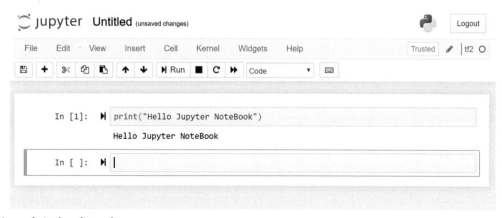

FIG. 14 Running code in the edit mode.

- Command mode: The mode for editing the Notebook, which can be identified by the cell with a gray border and blue left margin, as shown in Fig. 15. To enter the command mode, press the "Esc" key or click anywhere outside the cell.

FIG. 15   Command Mode.

2. Navigation in Jupyter Notebook
   - Mouse navigation: Using the mouse to perform any action such as selecting kernel, entering the modes, running selected cell, and so on from the interface of the Notebook.
   - Keyboard navigation: Using keyboard shortcuts to perform specific actions such as running the current cell, saving program, inserting cell above, copying selected cells, and so on. To see the keyboard shortcuts and their functions, refer to "Help → Keyboard Shortcuts" from the menu bar or click the "keyboard" icon from the tool bar, as shown in Fig. 16.

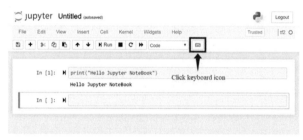

FIG. 16   Icon for displaying keyboard shortcuts.

Some keyboard shortcuts and their functions in Jupyter Notebook are shown in Fig. 17.

FIG. 17   Keyboard shortcuts in Jupyter Notebook.

5 PyCharm IDE

5.1 PyCharm installation

PyCharm is one of the most versatile integrated development environments (IDEs) for computer programing. It was developed by a Czech software development company and has many features, including coding assistance, project and code navigation, a graphical debugger, support for web frameworks, and so on. Therefore, it is recommended to use PyCharm for implementing complex programs. To download the PyCharm installation package, please go to the official website: www.jetbrains.com/pycharm/download, where there are professional and community versions of PyCharm for both Window and Ubuntu operating systems. The community version is free to use, while the professional version requires payment, as shown in Fig. 18. After downloading, please follow the installation steps to install.

Supplementary explanation

Remarks: If you are a student or teacher, you can go to www.jetbrains.com/student/ to apply for one-year free educational licenses to use the professional version of PyCharm. The licenses can be renewed free of charge if you are still a student or teacher after 1 year of use.

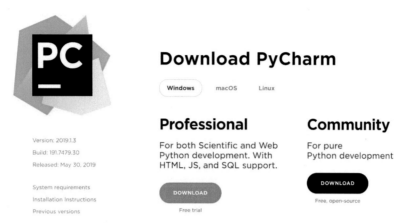

FIG. 18   Download PyCharm page.

5.2 Setup and create new project

After the installation of PyCharm, personal environment settings are required.

1. Set UI theme: This can be set as Darcula or Light, as shown in Fig. 19. The UI theme also can be changed later in Settings.

FIG. 19   Setting UI theme in PyCharm.

2. Install additional kits: PyCharm provides a few plugins in its repository for downloading and installing. If it is not necessary, this step can be skipped, as shown in Fig. 20.

FIG. 20 Downloading featured plugins in PyCharm.

3. Create a new project: Click "Create New Project" on the window of PyCharm, as shown in Fig. 21.

FIG. 21 Creating new project in PyCharm.

4. Select or create a directory for the project, as shown in Fig. 22.

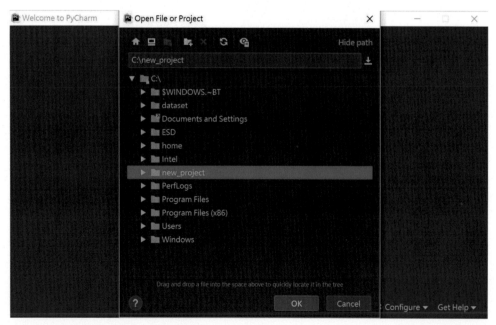

FIG. 22 Creating or selecting a directory for a new project.

5. Create a Python execution file: Right-click on project name → New → Python File, as shown in Fig. 23.

FIG. 23 Creating a new Python execution file in PyCharm.

6. Configure a Python interpreter: Click File → Settings → Project: Code → Project Interpreter → Add → Virtualenv Environment → Existing environment → Select the virtual machine such as tf2 or tf2-gpu (please refer to Section 2. TensorFlow Installation), as shown in Figs. 24 and 25.

FIG. 24 Project interpreter.

FIG. 25 Selecting an existing virtual environment.

7. Run the Python file: The Python file can be run by using the button icons from the top-right corner of the project window or by right-clicking on the Python file, as shown in Fig. 26.

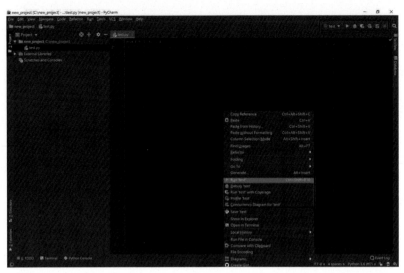

FIG. 26   Running the Python file.

5.3 PyCharm keyboard shortcuts

PyCharm provides keyboard shortcuts for quickly performing most of the tasks related to editing, debugging, navigation, and so on. This helps users work more effectively with keeping their hand on the keyboard. To see all keyboard shortcuts and their function in PyCharm, click "Help → Keymap Reference" from the menu bar, as shown in Figs. 27 and 28.

FIG. 27   Help menu in PyCharm.

PyCharm

Find any action inside the IDE   Ctrl + Shift + A

CREATE AND EDIT

Show intention actions	Alt + Enter
Basic code completion	Ctrl + Space
Smart code completion	Ctrl + Shift + Space
Type name completion	Ctrl + Alt + Space
Complete statement	Ctrl + Shift + Enter
Parameter information / context info	Ctrl + P / Alt + Q
Quick definition	Ctrl + Shift + I
Quick / external documentation	Ctrl + Q / Shift + F1
Generate code	Alt + Insert
Override / implement members	Ctrl + O / Ctrl + I
Surround with...	Ctrl + Alt + T
Comment with line comment	Ctrl + /
Extend / shrink selection	Ctrl + W / Ctrl + Shift + W
Optimize imports	Ctrl + Alt + O
Auto-indent lines	Ctrl + Alt + I
Cut / Copy / Paste	Ctrl + X / Ctrl + C / Ctrl + V
Copy document path	Ctrl + Shift + C
Paste from clipboard history	Ctrl + Shift + V
Duplicate current line or selection	Ctrl + D
Move line up / down	Ctrl + Shift + Up / Down
Delete line at caret	Ctrl + Y
Join / split line	Ctrl + Shift + J / Ctrl + Enter
Start new line	Shift + Enter
Toggle case	Ctrl + Shift + U
Expand / collapse code block	Ctrl + NumPad + / -
Expand / collapse all	Ctrl + Shift + NumPad + / -
Save all	Ctrl + S

VERSION CONTROL

MASTER YOUR IDE

Find action...	Ctrl + Shift + A
Open a tool window	Alt + [0-9]
Synchronize	Ctrl + Alt + Y
Quick switch scheme...	Ctrl + `
Settings...	Ctrl + Alt + S
Jump to source / navigation bar	F4 / Alt + Home
Jump to last tool window	F12
Hide active / all tool windows	Shift + Esc / Ctrl + Shift + F12
Go to next / previous editor tab	Alt + Right / Alt + Left
Go to editor (from a tool window)	Esc
Close active tab / window	Ctrl + Shift + F4 / Ctrl + F4

FIND EVERYTHING

Search everywhere	Double Shift
Find / replace	Ctrl + F / R
Find in path / Replace in path	Ctrl + Shift + F / R
Next / previous occurence	F3 / Shift + F3
Find word at caret	Ctrl + F3
Go to class / file	Ctrl + N / Ctrl + Shift + N
Go to file member	Ctrl + F12
Go to symbol	Ctrl + Alt + Shift + N

NAVIGATE FROM SYMBOLS

Declaration	Ctrl + B
Type declaration (JavaScript only)	Ctrl + Shift + B
Super method	Ctrl + U
Implementation(s)	Ctrl + Alt + B
Find usages / Find usages in file	Alt + F7 / Ctrl + F7
Highlight usages in file	Ctrl + Shift + F7
Show usages	Ctrl + Alt + F7

REFACTOR AND CLEAN UP

Refactor this...	Ctrl + Alt + Shift + T
Copy... / Move...	F5 / F6
Safe delete...	Alt + Delete

ANALYZE AND EXPLORE

Show error description	Ctrl + F1
Next / previous highlighted error	F2 / Shift + F2
Run inspection by name...	Ctrl + Alt + Shift + I
Type / call hierarchy	Ctrl + H / Ctrl + Alt + H

NAVIGATE IN CONTEXT

Select in...	Alt + F1
Recently viewed / Recent locations	Ctrl + E / Ctrl + Shift + E
Last edit location	Ctrl + Shift + Back
Navigate back / forward	Ctrl + Alt + Left / Right
Go to previous / next method	Alt + Up / Down
Go to line / column...	Ctrl + G
Go to code block end / start	Ctrl +] / [
Add to favorites	Alt + Shift + F
Toggle bookmark	F11
Toggle bookmark with mnemonic	Ctrl + F11
Go to numbered bookmark	Ctrl + [0-9]
Show bookmarks	Shift + F11

BUILD, RUN, AND DEBUG

Run context configuration	Ctrl + Shift + F10
Run / debug selected configuration	Alt + Shift + F10 / F9
Run / debug current configuration	Shift + F10 / F9
Step over / into	F8 / F7
Smart step into	Shift + F7
Step out	Shift + F8
Run to cursor / Force run to cursor	Alt + F9 / Ctrl + Alt + F9
Show execution point	Alt + F10
Evaluate expression...	Alt + F8
Stop	Ctrl + F2
Stop background processes...	Ctrl + Shift + F2
Resume program	F9
Toggle line breakpoint	Ctrl + F8
Toggle temporary line breakpoint	Ctrl + Alt + Shift + F8
Edit / view breakpoint	Ctrl + Shift + F8

FIG. 28   Keyboard shortcuts in PyCharm.

6 GitHub labs

6.1 Download source codes

The source codes of the Labs in the book are placed on GitHub. To download all the Labs, please go to GitHub URL: https://github.com/taipeitechmmslab/MMSLAB-DL/tree/master, then click "Code → Download ZIP" to download the Zip file, as shown in Fig. 29. After the download is completed, you must unzip this file for opening and running.

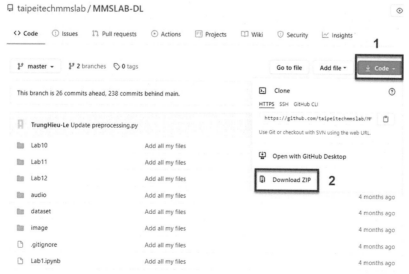

FIG. 29   The Labs of the book on GitHub.

6.2 Open and run source code

Suppose that the project directory path is C:\Users\mmslab\Deep-Learning-Book-master. To open the project, go to the folder "Deep-Learning-Book-master"

0. On Windows environment: Open the command prompt and run the following:

```
cd C:\Users\mmslab\Deep-Learning-Book-master
```

1. On Ubuntu environment: Open terminal and run the following:

```
cd /home/mmslab/Deep-Learning-Book-master
```

2. Open Jupyter Notebook: Run the following, the result is shown in Fig. 30:

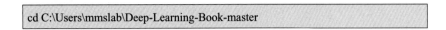

```
jupyter notebook
```

FIG. 30   Open the project through Jupyter Notebook.

3. To open the program of a specific Lab, click the corresponding file in the project. For example, to open the code of Lab1, please click the "Lab1.ipynb" file, as shown in Fig. 31.

FIG. 31   The source code of Lab1.

4. To run the program, click the "fast forward icon," as shown in Fig. 32.

FIG. 32   Running the code on Jupyter Notebook.

1

Introduction to TensorFlow 2

OUTLINE
- Getting to know deep learning
- Improvement of TensorFlow 2
- Getting started with TensorFlow 2
- Building neural networks using tf.keras
- Understanding of tf.data for building input pipelines

1.1 Deep learning

1.1.1 Introduction to deep learning

Artificial Intelligence (AI) refers to giving machines the ability to learn and mimic human actions. It is mainly used to assist humans, especially by replacing high-repetition, low-skill jobs. AI is a comprehensive field that includes evolutionary computation, expert systems, symbolic AI, support vector machines (SVMs), machine learning, deep learning, reinforcement learning, and many other fields.

Machine learning is a branch of AI that involves three main fields: reinforcement learning, deep learning, and SVMs, as shown in Fig. 1.1. Among them, deep learning is the most popular because it has been widely and successfully applied to various application areas such as computer vision, speech recognition, natural language processing, and so on [1]. Deep learning is based on artificial neural networks, also known as deep neural networks that have neural layers, namely, input layers, hidden layers, and output layers, and connections of neurons between these layers for representation learning. The "deep" in deep learning refers to the use of multiple layers through which the data can be transformed in the neural network.

In deep learning, the deep neural network is a complex function. When inputting a set of training data to the neural network, a set of output predictions can be obtained, and then the weights of the network will be updated to make the output predictions closer to the expected outputs, that is, the answers are marked.

1.1.2 Deep learning toolkits

Numerous open-source tools have been introduced to develop and train machine learning and deep learning models. The ten most popular deep learning toolkits are as follows:

1. TensorFlow: An end-to-end, open-source platform written in Python, C++ that provides a comprehensive flexible ecosystem of tools and libraries to help developers easily build and deploy machine learning- and deep learning-powered applications. More detail about TensorFlow is provided at the official website: http://tensorflow.org, GitHub: https://github.com/tensorflow/tensorflow.

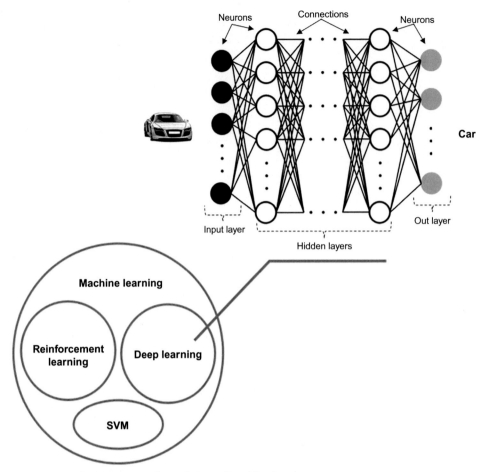

FIG. 1.1   Deep learning is currently the most popular technique of machine learning.

2. **Keras:** A high-level application programming interface (API) written in Python with the purpose of enabling fast experimentation. Keras can be run on top of other toolkits such as TensorFlow, Theano, and CNTK. More detail about Keras is provided at the official website: http://keras.io/, GitHub: https://github.com/keras-team/keras.

3. **Caffe:** A deep learning framework written in C++ developed by Berkeley Artificial Intelligence Research. Caffe is a flexible toolkit; models and optimization can be expressed by configuration without hard coding. More detail about Caffe is provided at the official website: http://caffe.berkeleyvision.org, GitHub: https://github.com/BVLC/caffe.

4. **MXNet:** An efficient library for deep learning that quickly trains network models and supports multiple programming languages, including Python, Scala, Clojure, Java, Julia, Perl, C++, and R. More detail about MXNet is provided at the official website: http://mxnet.io, GitHub: https://github.com/apache/incubator-mxnet.

5. **Theano:** A Python library for efficiently optimizing and evaluating mathematical expressions regarding multi-dimensional arrays. Symbolic differentiation can be efficiently performed through Theano with the use of graphics processing units (GPUs). More detail about Theano is provided at the official website: http://deeplearning.net/software/theano, GitHub: https://github.com/Theano/Theano.

6. The Microsoft Cognitive Toolkit (CNTK): By using CNTK for building deep learning models, the networks will be described as a series of computational steps through a directed graph. More detail about CNTK is provided at the official website: https://docs.microsoft.com/en-us/cognitive-toolkit, GitHub: https://github.com/Microsoft/CNTK.

7. DeepLearning4J (DL4J): A distributed deep learning library written in Java that is compatible with Java virtual machine languages, namely, Scala, Clojure, or Kotlin. DeepLearning4J is equal to Caffe in terms of performance when using multiple GPUs. More detail about DL4J is provided at the official website: https://deeplearning4j.org, GitHub: https://github.com/deeplearning4j/deeplearning4j.

8. PyTorch: A Python package that provides tensor computations like Numpy with sped-up GPU performance and that supports a neural network library with maximum flexibility. More detail about PyTorch is provided at the official website: http://pytorch.org, GitHub: https://github.com/pytorch/pytorch.

9. Chainer: A Python-based deep learning framework that supports building and training neural networks by providing automatic differentiation APIs and object-oriented high-level APIs. More detail about Chainer is provided at the official website: https://chainer.org, GitHub: https://github.com/chainer/chainer.

10. Deep Learning GPU Training System (DIGITS): A web application that supports frameworks including Caffe, TensorFlow, and Torch to accelerate training neural network models for image classification, segmentation, and object detection tasks. More detail about DIGITS is provided at the official website: https://developer.nvidia.com/digits, GitHub: https://github.com/NVIDIA/DIGITS.

The main purpose of building the above-mentioned toolkits is to provide a convenient framework for users to easily construct machine learning models in an efficient way. Compared to other toolkits, TensorFlow attracts the greatest number of users on GitHub with its top-score rating, which is calculated based on the number of GitHub contributions, Issues, Forks, and Stars [2]. Therefore, we chose TensorFlow to build and train deep learning models in the Labs in this book.

1.2 Introduction to TensorFlow

TensorFlow [3] is an end-to-end, open-source platform developed by the Google Brain team that uses data flow graphs to construct machine learning and deep learning models.

1. Tensor: refers to data representation

Tensor is an array of multi-dimensions with a uniform type called a *dtype*. All computations in TensorFlow are based on tensors. Some examples of creating basic tensors are listed here.

- Create a "scalar" tensor (a tensor with no "axes")

```
import tensorflow as tf
x1 = tf.constant(5)  # dtype=int32 (default)
print(x1) # Display x1
```

Result: tf.Tensor(5, shape=(), dtype=int32)

■ Create a "vector" tensor (a tensor has 1-axis):

| 1.0 | 2.0 | 6.0 | a vector, shape: [3]

```
import tensorflow as tf
x2 = tf.constant([1.0, 2.0, 6.0])  # dtype= float32
print(x2) # Display x2
```
Result: tf.Tensor([1. 2. 6.], shape=(3,), dtype=float32)

■ Create a "maxtrix" tensor (a tensor has 2-axes)

a matrix, shape: [3, 2]

```
import tensorflow as tf
x3 = tf.constant([[1, 8],
                  [2, 5],
                  [4, 4]], dtype=tf.float64)
print(x3) # Display x3
```
Result: tf.Tensor([[1. 8.]
                   [2. 5.]
                   [4. 4.]], shape=(3, 2), dtype=float64)

■ Create a 3-axes tensor with shape: [3, 2, 4]

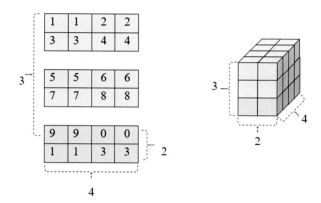

```
import tensorflow as tf
x4 = tf.constant ([[[1, 1, 2, 2],
                    [3, 3, 4, 4]],
                   [[5, 5, 6, 6],
                    [7, 7, 8, 8]],
                   [[9, 9, 0, 0],
                    [1, 1, 3, 3]],])
print(x4) # Display x4
```

Result: tf.Tensor([[[1 1 2 2]
[3 3 4 4]]

[[5 5 6 6]
[7 7 8 8]]

[[9 9 0 0]
[1 1 3 3]]], shape=(3, 2, 4), dtype=int32)

2. Flow: refers to a computational graph

TensorFlow works by using a data flow graph to represent computations. In the computational graph, each node represents an operation such as addition, subtraction, multiplication, division, and so on, and the edges represent the correlation between nodes. After each operation is performed, a new tensor will be formed for the next computation. For example, using TensorFlow to present the mathematical formula $Y = (a+b)/(a-b)$. The source code is as below:

```
import tensorflow as tf
a = tf.constant(4, name='a')   # a = 4
b = tf.constant(2, name='b')   # b = 2
c = tf.add (a, b, name='c')   # a + b
d = tf.subtract (a, b, name='c')   # a - b
Y = tf.divide (c, d, name='Y')   # c/d
print("Y =", Y) # Display Y
```

The computation graph for the above program is shown in Fig. 1.2. Note that TensorFlow uses TensorBoard to draw the graph; we introduce this function in Chapter 7. In Fig. 1.2, starting from the left, constants a and b are connected to the addition (Add) and subtraction (Sub) operation nodes to perform computations, then the resulting tensors from the Add node and Sub node are connected to the division (Div) operation node for computing the final result (Y).

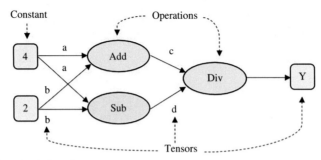

FIG. 1.2   Computational graph of $Y = (a+b)/(a-b)$.

In TensorFlow 1, a computational graph is first generated, and then a session is created to execute the graph. This technique is called a static computation graph. To help the user easily get started with TensorFlow and debug programs, Eager execution, an imperative programming environment has been introduced in TensorFlow 2 to perform operations immediately without constructing graphs; it also lessens boilerplate. In addition to Eager execution, there are many changes in TensorFlow 2, which we introduce in the next section.

1.3 Improvement of TensorFlow 2

TensorFlow 2 is an improved version of TensorFlow 1 with a focus on simple execution, easy model building, and robust model deployment. There are six main updated functions in TensorFlow 2: Eager Execution, Keras, tf.data, TensorFlow Hub, Distribution Strategy, and SavedModel. The first three functions are discussed in the next section. Later chapters discuss the other functions.

- Eager Execution: A flexible platform for research and experimentation on machine learning and deep learning, providing an imperative programming environment where operations can be evaluated immediately without constructing graphs.
- Keras: Keras was built on top of TensorFlow 2 as a high-level API for quickly and easily designing and training network models. Built-in Keras in TensorFlow 2 can be used through the "tf.keras" function.
- tf.data: An API that allows users to construct complex input pipelines in simple ways. This API can handle larger amounts of data and read data with different formats.
- TensorFlow Hub: A place where the trained machine learning and deep learning models are stored for reusing with a specific purpose such as fine-tuning a network model to address a new task with less training data and training time.
- Distribution Strategy: tf.distribute.Strategy, a TensorFlow API used to distribute the existing models for training across multiple devices, such as GPUs or TPUs.
- SavedModel: TensorFlow 2 has standardized a unifying format for storing complete network models with weights. Using a SavedModel, the trained models can be deployed to many platforms or devices, such as smartphones, Raspberry Pi, or even webpages.

Fig. 1.3 shows the simplified conceptual diagram of TensorFlow 2. As shown, the diagram consists of two stages: (1) training stage and (2) deployment stage. The training stage starts from reading and processing data with tf.data API for input pipelines, loading pre-trained model weights from TensorFlow Hub for fine-tuning, building neural networks with tf.keras API, training models with distribution strategy, and saving model weights with SavedModel. In the deployment stage, the trained model can be deployed in different devices or platforms through different libraries.

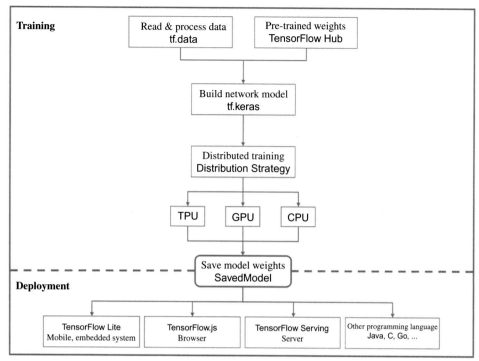

FIG. 1.3 The simplified conceptual diagram for TensorFlow 2.

- TensorFlow Lite: A library for converting the trained machine learning model into a compressed flat buffer and deploying it in mobile and Internet of Things (IoT) devices.
- TensorFlow.js: A library for deploying the trained machine learning model in JavaScript environments such as the browser or Node.js.
- TensorFlow Serving: A flexible serving system for machine learning and deep learning models, allowing for deployment of new algorithms while the server architecture remains unchanged.
- TensorFlow 2 also supports other programming languages such as Java, C, Go, and so on.

1.4 Eager execution

1.4.1 Introduction to eager execution

Immediate evaluation of operations through the Eager Execution platform is one of the most important updated features in TensorFlow 2. Unlike TensorFlow 1, which evaluates computations by first building a computational graph and then setting a session to execute the generated graph, Eager Execution in TensorFlow 2 provides a programming environment where the operations can be evaluated instantly without building graphs. This allows users to more easily debug with immediate error reporting as well as simplify the specification of dynamic models.

Important features of Eager Execution are as follows:

- Immediate evaluation of operations.
- Easier debugging program.
- No need to construct the computational graph.
- The calculation result can be returned without tf.Session.run.

Comparisons between TensorFlow 1 and TensorFlow 2:

- TensorFlow 1 code:
  - Define a constant: x = tf.constant(1), "x" does not execute.
  - Set a session: using tf.Session() to execute "x"

```
x = tf.constant(1)   # Create a scalar Tensor

print(x)   # display the constant Tensor information, shape=() means a scalar,
dtype=int32 represents an integer

# Create a session
sess = tf.Session()

# Use sess.run to execute the graph and display the value of the constant Tensor

print("x = {}".format(sess.run(x)))

# Close the Session
sess.close()
```

Result: Tensor("Const_5:0", shape=(), dtype=int32)
          x=1

- TensorFlow 2 code:
  - Define a constant: x = tf.constant(1), x will be evaluated immediately without setting and running any sessions.

```
x = tf.constant(1)   # Create a constant Tensor

print(x)   # display the constant Tensor, shape=() means it is a scalar, dtype=int32
represents an integer
```

Result: tf.Tensor(1, shape=(), dtype=int32)

1.4.2 Basic TensorFlow operations

This section presents a simple program, where the Eager Execution is first checked to ensure that it is activated. Then, we introduce and execute some basic operations such as addition, subtraction, multiplication, and so on to help readers understand a complete TensorFlow program.

- Step 1. Import necessary packages

```
import numpy as np  # import Numpy library

import tensorflow as tf  # import TensorFlow library

# check if Eager Execution is activated

print("Eager Execution is activated: {}".format(tf.executing_eagerly()))
```

Result: Eager Execution is activated: True

■ Step 2. Create a constant tensor

```
a = tf.constant(3)    # create a constant tensor with value of 3

b = tf.constant(4)    # create a constant tensor with value of 4

# display Tensor values

print("a = {}".format(a))

print("b = {}".format(b))
```

Result: a = 3

b = 4

■ Step 3. Check the data type of a tensor

```
print(a)   # shape=() means "a" is a scalar, and dtype=int32 means "a" is an integer.

print(b)   # shape=() means "b" is a scalar, and dtype=int32 means "b" is an integer.
```

Result: tf.Tensor(3, shape=(), dtype=int32)

tf.Tensor(4, shape=(), dtype=int32)

■ Step 4. Use the addition and multiplication operations

```
c = a + b

print("a + b = {}".format(c))    # show the result of a+b

d = a * b

print("a * b = {}".format(d))    # show the result of a*b
```

Result: a + b = 7

a * b = 12

■ Step 5. Create two-dimensional Tensors.

```
# Create a 2D Tensor whose dtype is float32

a = tf.constant([[1., 2.], [3., 4.]], dtype=tf.float32)

# Create a 2D Numpy array whose dtype is float32

b = np.array([[1., 0.], [2., 3.]], dtype=np.float32)

print("a constant: {}D Tensor".format(a.ndim))

c = a + b

print("a + b = \n{}".format(c))    # Display the result of a+b

# tf.matmul is matrix multiplication

d = tf.matmul(a, b)

print("a * b = \n{}".format(d))    # Display the result of a*b
```

Result: a constant: 2D Tensor

a + b = [[2. 2.]

[5. 7.]]

a * b = [[5. 6.]

[11. 12.]]

■ Step 6. Tensor can be converted to Numpy

```
print("NumpyArray:\n {}".format(c.numpy()))
```

Result: NumpyArray:
      [[2. 2.]
      [5. 7.]]

Supplementary explanation

The data format of TensorFlow can be different from that of other libraries, such as OpenCV or Matplotlib. When data type errors occur, the fastest solution is to convert data to the Numpy format since this format can be applied to most Python packages.

1.5 Keras

1.5.1 Introduction to Keras

Introduced by François Chollet in 2014, Keras [4] is an open-source, deep learning library designed for quickly building and training neural network models. Keras has no low-level operations of its own; it is run on top of open-source deep learning libraries called backend, such as TensorFlow, CNTK, and Theano. Keras utilizes Tensor-Flow backend by default; it can be changed to CNTK backend or Theano backend by using "use_backend()" function. There are a lot of commonly used neural-network building modules integrated into Keras for effectively designing models such as convolution layers, loss functions, activation functions, optimizers, and so on. In TensorFlow 2, Keras is the high-level API, so there is no need to install any additional Keras package; it can be used directly through the tf. keras API. It is highly recommended to use "tf.keras" API instead of "Keras" API because tf.keras supports Tensor-Flow operations, Eager Execution, tf.data, TPU training, and so on.

The following introduces the two most commonly used methods of building network models in Keras, including sequential models and functional models.

The following network layers are used for designing network models (later chapters present more detail about these layers).

- Dense: fully connected layer.
- Conv2d: convolution layer.
- Flatten: for flattening a tensor into a one-dimensional tensor. It is usually inserted between convolution layer and fully connected layer.
- Add: for adding a list of tensors with the same shape and outputting a single tensor.
- Concatenate: for concatenating a list of tensors; all tensors have the same shape except for the concatenation axis.

Supplementary explanation

Using **tensorflow.keras.utils.plot_model** requires additional installation of **pydot** and **graphviz** packages.

- On Ubuntu: Open Terminal and enter the following command:

pip install pydot
sudo apt install graphviz

- On Windows:

1. Open the command prompt and enter the following command:

pip install pydot

2. Go to https://graphviz.gitlab.io/_pages/Download/windows/graphviz-2.38.msi to Download and install the "graphviz" package, as shown in Fig. 1.4.

FIG. 1.4   Download Graphviz 2.38.

3. Finally, add the graphviz/bin installation path to "New system variable": Control Panel → System → Advanced System Settings → Environmental Variables, as shown in Fig. 1.5.

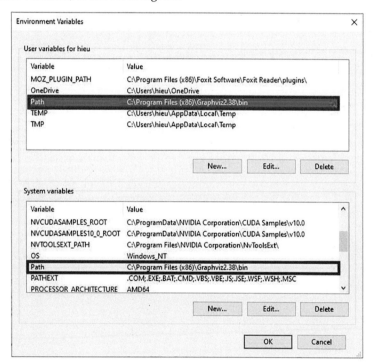

FIG. 1.5   New environment variables.

1.5.2 Sequential model

Sequential model is a method of building deep learning networks layer by layer in a systematic fashion. Fig. 1.6 shows a sequential network that starts from an input layer, which is connected to hidden layer 1. Then, hidden layer 1 is connected to hidden layer 2, and finally, hidden layer 2 and the output layer are connected to each other to form a complete model. Keras provides a sequential model API for constructing a network model easily and quickly. However, the construction method of the sequential model has some limitations.

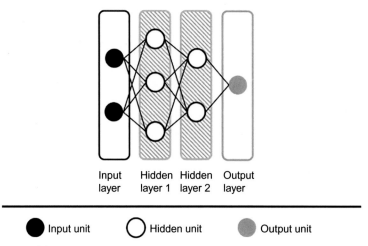

FIG. 1.6   Single input and output model.

- Only allows creating models with single input and single output.
- Does not allow creating models that share layers.
- Does not allow building models with multiple inputs or outputs.

Here, we present two methods of building a sequential network model, taking the image classification problem as an example. The model has an input with size of 28 × 28, which will be flattened into a one-dimensional vector of 784; two hidden layers (dense layers) with rectified linear (ReLU) activation functions; and an output followed by softmax layer for classification.

- Step 1. Import necessary packages

```
import tensorflow as tf

from tensorflow import keras

from tensorflow.keras import layers

from tensorflow.keras.utils import plot_model

from IPython.display import Image
```

- Step 2. Build the network model

The first method: Using "keras.Sequential"APT in Keras

```
# Create a Sequential model

model = keras.Sequential(name='Sequential')

# Each time model.add adds a layer to the network, the first layer needs to define the input size.

model.add(layers.Dense(64, activation='relu', input_shape=(784,)))

model.add(layers.Dense(64, activation='relu'))

# The last layer will be regarded as the output layer of the model

model.add(layers.Dense(10, activation='softmax'))
```

The second method: Using "tf.keras.Sequential" API in TensorFolow2.0

```
# all the network layers in a list and use them as parameters of tf.keras.Sequential

# And this list is also sequential, the first one needs to define the input size, and the last one is
the output layer

model = tf.keras.Sequential([layers.Dense(64, activation='relu', input_shape=(784,)),

                layers.Dense(64, activation='relu'),

                layers.Dense(10, activation='softmax')])
```

■ Step 3. Display the network model

```
# Generate network topology.

plot_model(model, to_file='Functional_API_Sequential_Model.png')

# Show the network topology.

Image('Functional_API_Sequential_Model.png')
```

Result:

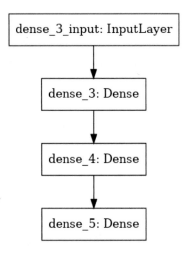

1.5.3 Functional model

The Keras functional API is a way to build network models by using a directed acyclic graph of layers. Following this method, instances of layers are first created, then they are connected to each other in pairs, and finally, the layers for acting as the input and output to the network are assigned. There are three steps for defining a network model using Keras functional API.

- Step 1 (defining input): Different from sequential model, the input layer with specific size of input data needs to be defined first.
- Step 2 (connecting layers): This step is completed by specifying where the input of a new layer comes from when it is declared.
- Step 3 (creating model): After all layers are declared and connected, it is required to specify the input and output of the model for completion of model creation.

The Keras functional API is very flexible for creating complex models, especially models with multiple inputs, multiple outputs, or multiple inputs and outputs, as shown in Fig. 1.7. All the complex models for advanced applications such as object detection, image segmentation, generative adversarial network, and so on, which we introduce later, can be built using a Keras functional API.

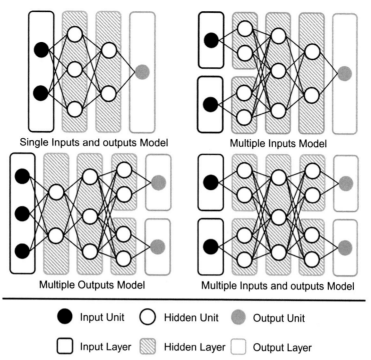

Single Inputs and outputs Model   Multiple Inputs Model

Multiple Outputs Model   Multiple Inputs and outputs Model

● Input Unit   ○ Hidden Unit   ● Output Unit

▢ Input Layer   ▨ Hidden Layer   ▢ Output Layer

FIG. 1.7 Different network model diagrams.

The following introduces how to build the four network models shown in Fig. 1.7 using a Keras functional API.

1. Single input and output model

For example, a model has an input with size of 28 × 28, which is flattened into a one-dimensional vector of 784; two hidden layers named "hidden1" and "hidden2" with ReLU activation functions; and an output layer named "Output," followed by softmax activation function for classification. The source code for constructing model is:

```
# Step1: Defining input layer

input = keras.Input(shape=(784,), name='Input')

#Step2: Connecting layers

#Hidden layers

h1 = layers.Dense(64, activation='relu', name='hidden1')(inputs)

h2 = layers.Dense(64, activation='relu', name='hidden2')(h1)

#Output layers

output = layers.Dense(10, activation='softmax', name='Output')(h2)

# Step 3: Creating network model

model = keras.Model(inputs=input, outputs=output)

plot_model(model, to_file='Functional_API_Single_Input_And_Output_Model.png')

Image('Functional_API_Single_Input_And_Output_Model.png')
```

Result:

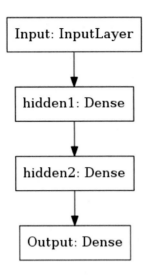

2. Multiple inputs and single output model

For example, a commodity price-prediction model consists of two inputs, including product image and product brand, and one output of price prediction with size of (1,). The input product image with size of (128, 128, 3) is passed through two hidden layers named "hidden1_1" and "hidden1_2", and the input product brand with size of (1,) goes through a hidden layer named "hidden1_3". After combining information of two inputs, the resulting features are sent to another hidden layer named "hidden2." Then, the output of "hidden2" layer is used as an input of a dense layer named "Output" for price prediction.

The source code for constructing this model is:

The source code for constructing model is as below:

```
# Step1: Defining Input layers

img_input = keras.Input(shape=(128, 128, 3), name='Image_Input')

info_input = keras.Input(shape=(1, ), name='Information_Input')

# Step 2: connecting layers

#Hidden layers

h1_1 = layers.Conv2D(64, 5, strides=2, activation='relu', name='hidden1_1')(img_input)

h1_2 = layers.Conv2D(32, 5, strides=2, activation='relu', name='hidden1_2')(h1_1)

h1_2_ft = layers.Flatten()(h1_2)

h1_3 = layers.Dense(64, activation='relu', name='hidden1_3')(info_input)

concat = layers.Concatenate()([h1_2_ft, h1_3])

h2 = layers.Dense(64, activation='relu', name='hidden2')(concat)

# Output layer
```

```
outputs = layers.Dense(1, name='Output')(h2)

# Step 3: Creating network model

model = keras.Model(inputs=[img_input, info_input], outputs=outputs)

# Show network model architecture

plot_model(model, to_file='Functional_API_Multi_Input_Model.png')

Image('Functional_API_Multi_Input_Model.png')
```

Result:

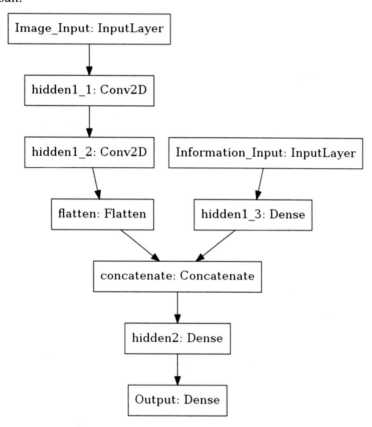

3. Single input and multiple outputs model

For example, a portrait-recognition model has an input and two outputs. The input of the model is a personal photo with size of (128, 128, 3), which is passed through three hidden layers named "hidden1," "hidden2," and "hidden3." After flattening the output of the last hidden layer (hidden3), it is sent to two output layers named "Age_Output" and "Gender_Output" for age and gender predictions, respectively.

The source code for constructing model is:

```
# Step1: Defining Input layers

inputs = keras.Input(shape=(128, 128, 3), name='Input')

#Step 2: connecting layers

# Hidden layers

h1 = layers.Conv2D(64, 3, activation='relu', name='hidden1')(inputs)

h2 = layers.Conv2D(64, 3, strides=2, activation='relu', name='hidden2')(h1)

h3 = layers.Conv2D(64, 3, strides=2, activation='relu', name='hidden3')(h2)

# flattening

flatten = layers.Flatten()(h3)

# Output layers

age_output = layers.Dense(1, name='Age_Output')(flatten)

gender_output = layers.Dense(1, name='Gender_Output')(flatten)

# Step 3: Creating network model

model = keras.Model(inputs=inputs, outputs=[age_output, gender_output])

# Show network model architecture

plot_model(model, to_file='Functional_API_Multi_Output_Model.png')

Image('Functional_API_Multi_Output_Model.png')
```

Result:

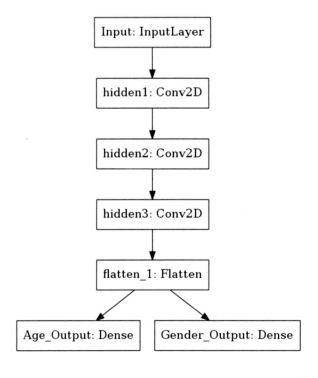

4. Multiple inputs and multiple outputs model

For example, a weather prediction model has two inputs for satellite cloud image and climate information, and three outputs for rainfall probability, temperature, and humidity. The satellite cloud image with size of (256x256x3) is passed through three hidden layers named "hidden1," "hidden2," and "hidden3," and climate information with size of (10,) is put through a dense layer. After combining two inputs of information, the model outputs climate information including rainfall probability, temperature, and humidity through "Output1," "Output2," and "Output3" layers, respectively.

The source code for constructing model is:

```
# Step1: Defining Input layers

image_inputs = keras.Input(shape=(256, 256, 3), name='Image_Input')

info_inputs = keras.Input(shape=(10, ), name='Info_Input')

#Step 2: connecting layers

#Hidden layers

h1 = layers.Conv2D(64, 3, activation='relu', name='hidden1')(image_inputs)

h2 = layers.Conv2D(64, 3, strides=2, activation='relu', name='hidden2')(h1)

h3 = layers.Conv2D(64, 3, strides=2, activation='relu', name='hidden3')(h2)
```

```
flatten = layers.Flatten()(h3)
# Dense layer
h4 = layers.Dense(64)(info_inputs)
concat = layers.Concatenate()([flatten, h4])
# Output layers
weather_outputs = layers.Dense(1, name='Output1')(concat)
temp_outputs = layers.Dense(1, name='Output2')(concat)
humidity_outputs = layers.Dense(1, name='Output3')(concat)
# Step 3: Creating network model
model = keras.Model(inputs=[image_inputs, info_inputs],
                outputs=[weather_outputs, temp_outputs, humidity_outputs])
# Show network model architecture
plot_model(model, to_file='Functional_API_Multi_Input_And_Output_Model.png')
Image('Functional_API_Multi_Input_And_Output_Model.png')
```

Result:

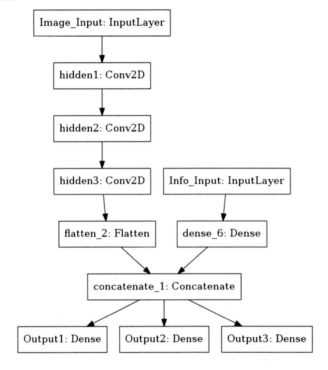

1.6 Tf.data

1.6.1 Introduction to tf.data

Preparing and feeding data into the model are the first steps in training machine learning and deep learning models. If the method of building a data input pipeline causes the CPU to take a long time to load data, and thus the GPU needs to wait for the CPU, this can lead to performance problems. To keep the GPU from data starvation, TensorFlow provides a "tf.data" API to facilitate building complex and highly optimized data pipelines. As shown in Fig. 1.8, before data are sent to the machine learning or deep learning model, they are processed with "tf.data" through three stages: extraction stage, transformation stage, and loading stage.

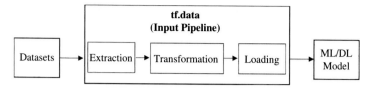

FIG. 1.8 Processing input data pipeline with "tf.data" API. Note that, ML and DL are abbreviation for machine learning and deep learning, respectively.

- Extraction: Reading data from storage such as local disks (SSD, HDD), cloud storage, and so on, is the first step in a data pipeline, and "tf.data" supports reading a lot of data formats. Fig. 1.9 is an illustration of reading and representing data with tf.data. Note that if data in this step are directly passed to the model, one of the rows of represented data will be taken in each training iteration. A dataset with size 5 TB and a CPU run with 16 GB of RAM will be trouble in the training model.

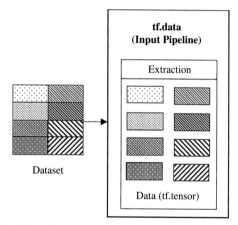

FIG. 1.9   Reading and representing data with "tf.data."

- Transformation: After reading and representing data, the common next step is to shuffle data and create mini-batches of data. "tf.data" uses "tf.data.Dataset" API to complete this task, as shown in Figs. 1.10 and 1.11.

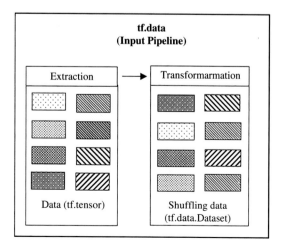

FIG. 1.10   Shuffling data with tf.data.

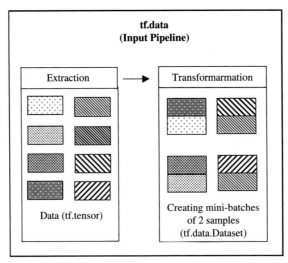

FIG. 1.11 Creating mini-batches of data with tf.data.

- Loading: After the transformation step, data is loaded into acceleration devices such as GPUs or TPUs for training models. The tf.data API provides prefetching of data, parallel input/output (I/O), and parallel processing of batches for high-performance load, as shown in Fig. 1.12.

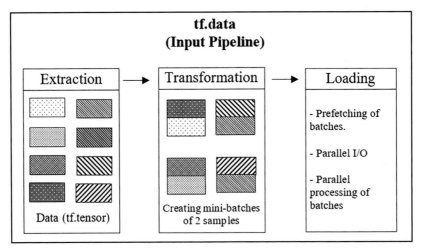

FIG. 1.12 Loading data with tf.data.

1.6.2 Basic functions of tf.data API

This section provides some example programs that use the tf.data API for preparing data such as creating data, setting data attributes, retrieving data, and so on. Note that it is necessary to execute each line of the code in order, starting from 1.

1. tf.data.Dataset.from_tensors: creating dataset with single element

```
dataset = tf.data.Dataset.from_tensors(tf.constant([1, 2, 3, 4, 5, 6, 7, 8, 9, 10], shape=(10, )))

# Display Data information: shapes and data type

print(dataset)
```

Result:<TensorDataset shapes: (10,), types: tf.int32>

2. tf.data.Dataset.from_tensor_slices: creating a dataset that whose elements are slices of tensors

```
x_data = tf.data.Dataset.from_tensor_slices(tf.constant([0, 1, 2, 3, 4, 5, 6, 7, 8, 9]))
# Display Data information: shapes and data type
print(x_data)
y_data = tf.data.Dataset.from_tensor_slices(tf.constant([0, 2, 4, 6, 8, 10, 12, 14, 16, 18]))
# Display Data information: shapes and data type
print(y_data)
```

Result: <TensorSliceDataset shapes: (), types: tf.int32>

<TensorSliceDataset shapes: (), types: tf.int32>

3. for loop: using for iterating over a sequence such as a list, a tuple, a string, and so on.

```
for dataset in dataset:
    print(data)
```

Result: tf.Tensor([1 2 3 4 5 6 7 8 9 10], shape=(10,), dtype=int32)

```
for data1, data2 in zip(x_data, y_data):
    print('x: {}, y: {}'.format(data1, data2))
```

Result: x: 0, y: 0

x: 1, y: 2

x: 2, y: 4

x: 3, y: 6

x: 4, y: 8

x: 5, y: 10

x: 6, y: 12

x: 7, y: 14

x: 8, y: 16

x: 9, y: 18

4. take: creating a dataset with count elements from the source dataset

```
for data in dataset.take(1):
    print(data)
```

Result: tf.Tensor([1 2 3 4 5 6 7 8 9 10], shape=(10,), dtype=int32)

```
for data1, data2 in zip(x_data.take(5), y_data.take(5)):
    print('x: {}, y: {}'.format(data1, data2))
```

Result: x: 0, y: 0

x: 1, y: 2

x: 2, y: 4

x: 3, y: 6

x: 4, y: 8

If the element count exceeds the number of elements of the source dataset, the new dataset is the source dataset.

```
for data1, data2 in zip(x_data.take(12), y_data.take(12)):
print('x: {}, y: {}'.format(data1, data2))
```

Result: x: 0, y: 0
      x: 1, y: 2
      x: 2, y: 4
      x: 3, y: 6
      x: 4, y: 8
      x: 5, y: 10
      x: 6, y: 12
      x: 7, y: 14
      x: 8, y: 16
      x: 9, y: 18

5. tf.data.Dataset.zip: creating a new dataset by zipping together the given datasets

```
tf.data.Dataset.zip((x_data, y_data))
```

Result:\<ZipDataset shapes: ((), ()), types: (tf.int32, tf.int32)>

6. map: transformation of data by using a specified function

```
tf.data.Dataset.range(10).map(lambda x: x*2)
```

Result: \<MapDataset shapes: (), types: tf.int64>

7. Name: the samples from dataset can be named

```
x = tf.data.Dataset.range(10)

y = tf.data.Dataset.range(10).map(lambda x: x*2)

dataset = tf.data.Dataset.zip({"x": x, "y": y})

print(dataset)
```

Result: \<ZipDataset shapes: {y: (), x: ()}, types: {y: tf.int64, x: tf.int64}>

```
for data in dataset.take(10):
    print('x: {}, y: {}'.format(data['x'], data['y']))
```

Result: x: 0, y: 0
      x: 1, y: 2
      x: 2, y: 4
      x: 3, y: 6
      x: 4, y: 8
      x: 5, y: 10
      x: 6, y: 12
      x: 7, y: 14
      x: 8, y: 16
      x: 9, y: 18

8. Set a size for each batch:

```
dataset = tf.data.Dataset.zip({"x": x, "y": y}).batch(2)

for data in dataset.take(5):

    print('x: {}, y: {}'.format(data['x'], data['y']))
```

Result: x: [0 1], y: [0 2]

x: [2 3], y: [4 6]

x: [4 5], y: [8 10]

x: [6 7], y: [12 14]

x: [8 9], y: [16 18]

9. Shuffle: Randomly shuffling the elements of the dataset. A buffer is filled with "buffer_size" elements, then these elements in this buffer are randomly sampled, and the selected elements are replaced with new elements. It is suggested to set the "buffer_size" to be greater than or equal to the size of the dataset.

```
dataset = dataset.shuffle(10)

for data in dataset.take(5):

    print('x: {}, y: {}'.format(data['x'], data['y']))
```

Result: x: [0 1], y: [0 2]

x: [6 7], y: [12 14]

x: [4 5], y: [8 10]

x: [8 9], y: [16 18]

x: [2 3], y: [4 6]

10. Repeat: After finishing reading all the samples from a dataset, no samples can be read unless a "repeat" function is used. Setting repeat(n) allows n times dataset to be repeated

```
for data in dataset.take(10):

    print('x: {}, y: {}'.format(data['x'], data['y']))

print('-' * 50)

dataset = dataset.repeat(2)

for data in dataset.take(10):
```

```
print('x: {}, y: {}'.format(data['x'], data['y']))
```

Result: x: [0 1], y: [0 2]

x: [6 7], y: [12 14]

x: [4 5], y: [8 10]

x: [8 9], y: [16 18]

x: [2 3], y: [4 6]

\---

x: [6 7], y: [12 14]

x: [0 1], y: [0 2]

x: [4 5], y: [8 10]

x: [8 9], y: [16 18]

x: [2 3], y: [4 6]

x: [8 9], y: [16 18]

x: [4 5], y: [8 10]

x: [6 7], y: [12 14]

x: [0 1], y: [0 2]

x: [2 3], y: [4 6]

References

[1] S. Dargan, M. Kumar, M.R. Ayyagari, G. Kumar, A survey of deep learning and its applications: a new paradigm to machine learning. Arch. Computat. Methods Eng. 27 (2020) 1071–1092, https://doi.org/10.1007/s11831-019-09344-w.

[2] J. Zacharias, M. Barz, D. Sonntag, A survey on deep learning toolkits and libraries for intelligent user interfaces, CoRR, 2018. vol. abs/1803.04818. [Online]. Available:http://arxiv.org/abs/1803.04818.

[3] M. Abadi, A. Agarwal, P. Barham, E. Brevdo, Z. Chen, C. Citro, G.S. Corrado, A. Davis, J. Dean, M. Devin, S. Ghemawat, I. Goodfellow, A. Harp, G. Irving, M. Isard, Y. Jia, R. Jozefowicz, L. Kaiser, M. Kudlur, J. Levenberg, D. Mane, R. Monga, S. Moore, D. Murray, C. Olah, M. Schuster, J. Shlens, B. Steiner, I. Sutskever, K. Talwar, P. Tucker, V. Vanhoucke, V. Vasudevan, F. Viegas, O. Vinyals, P. Warden, M. Wattenberg, M. Wicke, Y. Yu, X. Zheng, TensorFlow: Large-scale machine learning on heterogeneous systems, software available from tensorflow.org. [Online]. Available:https://www.tensorflow.org/, 2015.

[4] F. Chollet, Keras, https://github.com/fchollet/keras, 2015.

2

Neural networks

2.1 Introduction to neural networks

2.1.1 A brief history of neural networks

In 1943, Warren McCulloch and Walter Pitts [1] introduced a computational model based on a threshold logic algorithm, paving the way for the development of artificial neural networks (ANNs). Inspired from McCullock and Pitts [1], numerous methods have been proposed for neural networks which can be categorized into two approaches: (1) focusing on biological process [2,3] and (2) focusing on application [4,5]. However, around 1980–2000, simpler models such as linear classifiers or support vector machines (SVMs) [6,7] for classification and regression analysis became more popular. The reason can be explained as follows. Deep networks with many intermediate layers called hidden layers can obtain better performance than that of shallow neural networks with few hidden layers, as shown in Fig. 2.1. In a deep network, the hidden layers extract features from input data and these features are used on the following layers for computation. Therefore, the more hidden layers, the more detailed features of data can be obtained. For example, a human face recognition model takes face images as inputs, and then the first hidden layer takes raw data from the input layer and extracts simple features such as edges, lines, and so on. In the following hidden layers, the high-level specific features of the input image such as nose, eyes, hair, and others can be extracted for face recognition. Unfortunately, machine learning algorithms like gradient-based learning methods and backpropagation [8] do not work well for deep neural networks because of a vanishing gradient problem during the training process. Furthermore, since deep neural networks have high computational cost and memory footprint, they require large datasets for training and development.

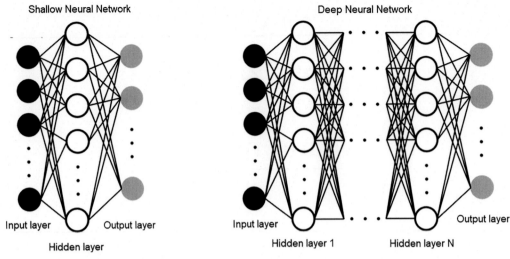

FIG. 2.1   The architecture of a neural network.

From 2000 to today, the development of various aspects of computer technology and machine learning algorithms has made it possible to build and train deep neural networks, and there have been major breakthroughs.

- Computing ability

In the past, central processing units (CPUs) were used to perform the calculation of neural networks, but now graphic processing units (GPUs) and tensor processing units (TPUs) are used because of their improved performance. In 2012, Krizhevsky et al. used a NVIDIA GPU to train an eight-layer network called "AlexNet" [9]. After winning the 2012 ImageNet Large Scale Visual Recognition Challenge (ILSVRC) competition [10], everyone realized the amazing computing power of GPUs and the abilities of deep neural networks.

- Dataset and storage equipment

A deep neural network cannot be trained efficiently without sufficient training data. Because data storage devices were expensive and had limited capacity, large datasets were impossible 20–30 years ago. Nowadays, with the development of science and technology, the price of storage devices has decreased, and Internet and cloud devices, such as Kaggle (www.kaggle.com/) and TensorFlow hub (www.tensorflow.org/hub) are prevalent; there are many available databases for users to download, research, and develop. Thus the problems of lacking of large datasets and data storage devices have been solved.

- Activation function

Sigmoid was a very popular activation function for neural networks. However, using the sigmoid activation function in hidden layers of deep networks with random weight initialization leads to vanishing gradients during training, which makes the weights of the networks difficult to update, resulting in ineffective training. In 2011, Glorot et al. [11] introduced the rectified linear unit (ReLU) function, which has proved effective at improving the problem of vanishing gradients and the learning speed of various deep neural networks. In 2015, parametric rectified linear unit (PReLU) activation was introduced by He et al. [12], allowing for effectively investigating and training deeper and wider deep neural network models.

2.1.2 Principle of neural networks

A neural network is composed of three types of layers: input layer, hidden layer, and output layer. Each layer in the network has many neurons for connection and computation, as shown in Fig. 2.2. The neurons of the input layer, hidden layer, and output layer are called the input neuron, hidden neuron, and output neuron, respectively. The input layer provides initial data from outside to the network without any computations for further processing by hidden layers. Hidden layers are located between the input and output layers in the neural network and are responsible for performing computation on the input data through hidden neurons and passing the results to the output layer. The output layer takes the results from the last hidden layer as the input data and uses its neurons to compute and produce the final result of the network.

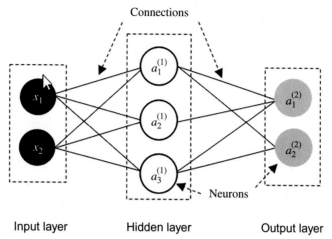

FIG. 2.2 Example of a two-layer neural network. Note that the number of layers of the neural network is equal to the number of hidden layers plus one, $a_i^{(l)}$ represents the output of i-th neuron in l-th layer, the biases are hidden.

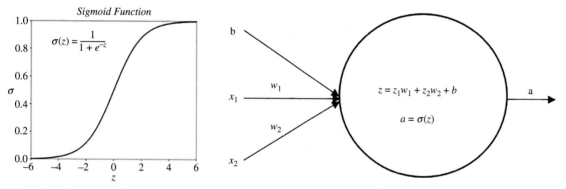

FIG. 2.3 Illustration of a neuron of a neural network. x_i, w_i, and b represent inputs, weights, and bias of the neuron, respectively. σ is an activation function.

Except for input neurons, each neuron of the neural network can receive one or more inputs with separate weights, and these inputs are summed to produce an output. Normally, after summing the inputs, the resulting sum is sent to a non-linear function, which is known as the activation function for generating output, as shown in Fig. 2.3. The common activation functions include ReLU, sigmoid, and tanh, which we discuss in later chapters.

When the neural network adopts activation functions to perform a nonlinear transformation of the inputs, it can help the neural networks describe and learn complex tasks.

2.1.3 Training neural networks

Training a neural network is the process of using training data to find the appropriate weights of the network for creating a good mapping of inputs and outputs. As shown in Fig. 2.4, the training procedure for a neural network consists of four parts: preparing the dataset, building a network model, loss function, and optimization.

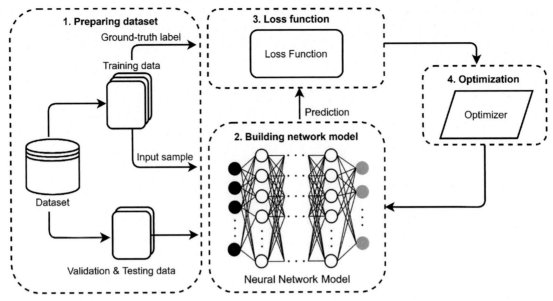

FIG. 2.4 Schematic diagram of the neural network training procedure.

1. Preparing the dataset

The dataset for training neural networks is divided into three types of data: training data, validation data, and test data.

- Training data: A set of examples, which is used for fitting the weights of connections between neurons in neural networks. The training data often contains pairs of samples (input sample, ground truth label). The ground truth label is also called the expected output.
- Validation data: A set of examples, which is used to estimate the model fit during tuning the hyperparameters of the network model, such as the number of hidden layers, the number of neurons in each layer, and so on.
- Test data: A set of examples, which is used to estimate the final network model fit on the training data. For most competitions, training data and validation data are published, while the test data is kept as the basis for the final evaluation of the model.

Supplementary explanation

The test data is used to evaluate the accuracy and generalization ability of a neural network in the real world. If the hyperparameters of the neural network are directly tuned through the results on the test data, the network can often obtain good performance on both training data and test data, but it will perform poorly when applied to the real world [13]. In order to avoid this problem, the validation data is used instead of the test set, so that the performance of the model on the test data will be more in line with that of the model operating in the real world.

2. Building a network model

As introduced in Section 2.1.2, neural networks consist of three main components: input layers, hidden layers, and output layers. For different tasks, the network models are built with different hyperparameters, such as the number of hidden layers, the number of neurons in each layer, and so on. Because the choice of hyperparameters directly affects the training results of the network models, we provide the TensorBoard tuning toolkit in Chapter 7 to assist in adjusting these hyperparameters.

In the neural network, if each neuron in a layer is connected to all neurons in the following layer, this network is called a fully connected neural network (FCNN). Fig. 2.5 shows a FCNN with three hidden layers that employ the ReLU activation function for non-linear transformation. Because of full connection, all the combinations of the information from the previous layer can be used to compute in the next layer; this helps the FCNN learn better input data

[14]. However, FCNNs are often extremely computationally expensive and are challenged by the problem of overfitting, a phenomenon of modeling the training data too well during the training process, which leads to poor performance of the network, especially deep neural networks [15]. In the next sections of this chapter, we discuss the application of an FCNN for house price-prediction, as well as provide the solution to the overfitting problem.

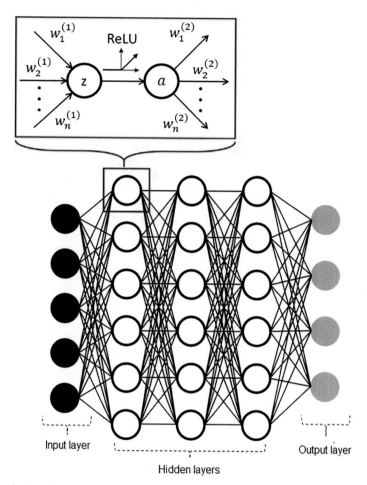

FIG. 2.5 Fully connected neural network.

3. Loss function

A neural network is trained through an optimization process that uses a loss function to calculate an error between the predicted value of the model and the expected output. For the different purposes of training, the optimization process may minimize or maximize the loss function, which means it needs to evaluate a suitable solution such as a set of parameters to reach the lowest or highest error score, respectively. Typically, minimizing the loss function is applied when training the neural networks. There are many loss functions and it can be challenging to choose a suitable one for a specific problem. The following introduces commonly used loss functions for main problems in machine learning, including the linear regression problem, binary classification problem, and multiclass classification problem.

- Linear regression problem: Neural networks are designed with one neuron in the output layer for each possible desired value. An example of a regression model is the house price-prediction model, which is based on information such as the number of bedrooms, bathrooms, and floors in the house, the age of the house, and so on, to predict the price of a house. The most common loss functions for linear regression models are mean squared error (MSE) and mean absolute error (MAE), which calculate the average of the squared or absolute difference between the predicted value and the expected output.
- Binary classification problem: Neural networks are designed with one neuron in the output layer for predicting an input sample as belonging to one of two classes. Medical testing to determine whether a person has a certain disease or not is a typical binary classification problem. For training binary classification models, binary cross-entropy loss function, also known as log loss function, is commonly used. We discuss this in more detail in Chapter 3.

- Multiclass classification problem: Neural networks are designed with many neurons in the output layers, where each neuron with softmax activation functions is responsible for predicting one class. Categorical cross-entropy loss function, also known as softmax loss function, is widely used for training multiclass classification models. We discuss categorical cross-entropy in more detail in Chapter 4.

Because house price-prediction introduced in this chapter is a linear regression problem, MSE or MAE can be employed as loss functions. Here, we present the formulas and differences of these two loss functions.
MSE:

$$MSE = \frac{\sum_{i=1}^{N}(y_i - \hat{y}_i)^2}{N}$$

MAE:

$$MAE = \frac{\sum_{i=1}^{N}|y_i - \hat{y}_i|}{N}$$

where y is the expected output, $\hat{y}$ is the predicted value of the neural network model, and N is the amount of data in a batch.

Both MSE and MAE calculate the error between the predicted value (y_pre) of the model and the expected output (y_true), but MSE calculates the average of squared error and MAE calculates the average of absolute error. Fig. 2.6 presents a comparison between MSE and MAE. As shown, when the value of y_true - y_pre is between -1 and 1, the MSE loss value is lesser than that of MAE. In contrast, the MSE loss value is greater than MAE loss value when the value of y_true - y_pre is greater than 1 or less than -1. Using these two methods as the loss function will produce different training results.

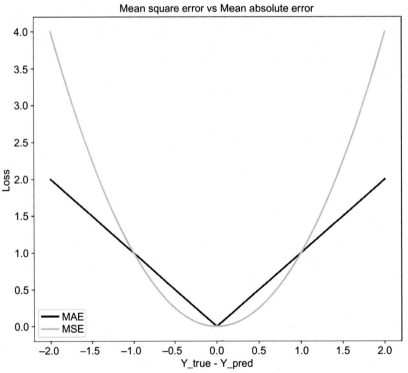

FIG. 2.6   MSE and MAE loss value.

The house price-prediction model in this chapter uses MSE as the training loss function. The reader can change MSE to MAE to compare the results between these two methods.

4. Optimization

In training neural networks, the optimization process is to find a set of parameters, also called a set of weights, to make the loss value of the loss function as small as possible. Gradient descent (GD) is the most commonly used optimization algorithm for training neural networks. To find a minimum of the loss function with GD, take steps proportional to the opposite direction of the gradient of the loss function at the current point. This algorithm can be likened to a person who is looking for a path to get down a mountain where the path down is not visible. Based on the current position of the person, they look at the steepness of the mountain, then go downhill in the direction of the steepest descent. This step is repeated until they reach the bottom of the mountain. For example, given a neural network model with uninitialized weights W and a loss function L which is used to calculate the error between the predicted output of the network and the expected output. The network is trained using the GD algorithm to minimize L. The steps of finding W to reach a minimum loss value of L are as follows:

- Step 1. Randomly initialize W
- Step 2. Update W with a learning rate η through the formula:

$$W = W - \eta \frac{\partial L}{\partial W}$$

where $\frac{\partial L}{\partial W}$ represents the gradient of L at point W.

- Step 3. Calculate the value of L with new W, called "loss," and repeat Step 2 until reaching a minimum loss, as shown in Fig. 2.7

The loss value reaches minimum loss, which means the predicted results of the model are closest to the expected outputs.

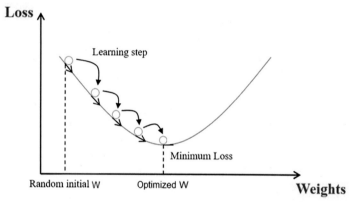

FIG. 2.7 Gradient descent diagram.

In fact, not every update of GD is updated towards the minimum value of the loss function, but rather is updated towards the direction that can reduce the error of the loss function at that time. Thus, when the model is trained until the loss value cannot be reduced, the reached point is usually the local minimum instead of the global minimum, as shown in Fig. 2.8.

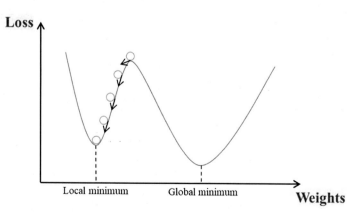

FIG. 2.8 The weights are updated in the direction of converging to a local minimum instead of a global minimum.

Learning rate is a configurable hyperparameter that determines step size at each iteration in training neural networks. A smaller learning rate results in small changes in the weights of each update and requires many training iterations, while a larger learning rate makes quick changes and requires fewer training iterations. If the learning rate is too large, the changes in the weights of each update are too large; it is likely to jump over the minimum value and produce oscillations, as shown in Fig. 2.9A. Conversely, if the learning rate is too small, the optimization efficiency may be poor, and the optimal value cannot be found after a long training, as shown in Fig. 2.9B. Thus, the learning rate is one of the most crucial hyperparameters to be carefully selected when training the network model.

FIG. 2.9 Optimization process with different learning rates.

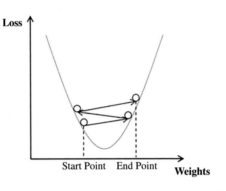

(A) Too large learning rate.

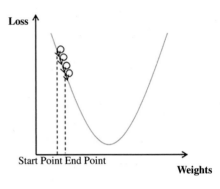

(B) Too small learning rate.

There are many kinds of gradient-based optimization algorithms. The aforementioned GD uses all training data to calculate the gradient of the loss function and update the weights once. If the neural network is updated N times, it needs to calculate the entire training data N times. Using GD is very time-consuming and inefficient. Therefore, the stochastic gradient descent (SGD) algorithm is introduced. At each time, SDG calculates the gradient of the loss functions based only on a random sample from the training dataset, and then updates the weights based on this gradient. This makes SDG suitable for training a huge training dataset. In addition, there are also many optimization methods, such as Momentum [16], AdaGrad [17], Adam [18], and others. Momentum adapts the concept of momentum while AdaGrad adjusts the learning rate according to the gradient for optimization. Adam can be seen as a combination of Momentum and AdaGrad that utilizes estimations of first and second moments of gradient to adapt learning rates for different weights. By using Adam as an optimizer, the weights of the network model can be updated through the following formula:

$$w_t = w_{t-1} - \alpha \frac{\hat{m}_t}{\sqrt{\hat{v}_t} + \varepsilon}$$

where w is the weights of the model, α is step size parameter ($\alpha = 0.001$), ε is set to 10^{-8}, $t = 0$ is for initialization of time step, and $\hat{m}_{t+1}$ and $\hat{v}_{t+1}$ are defined as:

$$\hat{m}_t = \frac{m_t}{1 - \beta_1^t}$$

$$\hat{v}_t = \frac{v_t}{1 - \beta_2^t}$$

Here, $\beta_1, \beta_2 \in [0, 1)$ are hyperparameters for controlling the exponential decay rates of the moving average of the gradient m_t and the squared gradient v_t.

$$m_t = \beta_1 m_{t-1} + g_t(1 - \beta_1)$$
$$v_t = \beta_2 v_{t-1} + (g_t \odot g_t)(1 - \beta_2)$$

where g_t represents gradient on mini-batch at timestep t, and $\odot$ is the element-wise operation.

Because of the efficiency in optimization, Adam has been widely applied for training deep neural networks in recent years.

2.2 Introduction to Kaggle

2.2.1 Kaggle platform

Kaggle is a world-famous competition platform where companies and researchers can publish datasets and organize competitions to solve the challenges of data science. The competition is open to anyone. As shown in Fig. 2.10, the Kaggle contest page has as many as 19 contests in progress. It is also worth noting that Kaggle is not only a competition website but also a community platform, allowing users to work, discuss, team up, or share research results with each other.

Another great thing about Kaggle is that it has a datasets area, in which datasets have been sorted and made available for download, as shown in Fig. 2.11. Experimental data in this chapter is a dataset downloaded from the Kaggle website.

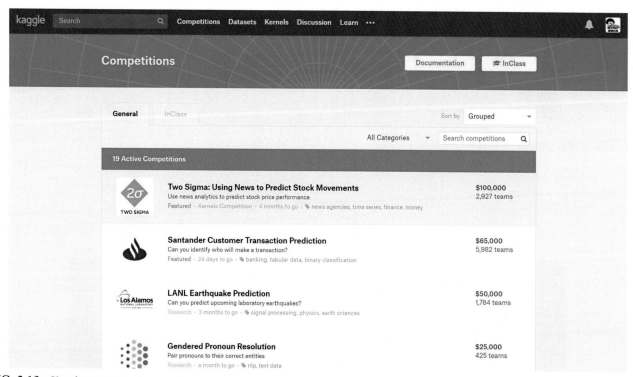

FIG. 2.10 Kaggle competition page.

2.2.2 House sales in King County dataset

In this section, a "House Sales in King County, USA" dataset from Kaggle is introduced for training and evaluating the house price-prediction model. To download the dataset, please access the URL:https://www.kaggle.com/harlfoxem/housesalesprediction, as shown in Fig. 2.12.

This dataset has 21,613 housing data, and each house sample has 21 items of information. The codes indicate the following meanings:

- id: identification code of the house
- date: date the house was sold
- price: housing price (target)
- bedrooms: number of bedrooms
- bathrooms: number of bathrooms
- sqft_living: area of the interior living space (square feet)
- sqft_lot: area of the land space (square feet)
- floors: total floors of the house
- waterfront: a variable for whether or not the apartment overlooks the waterfront

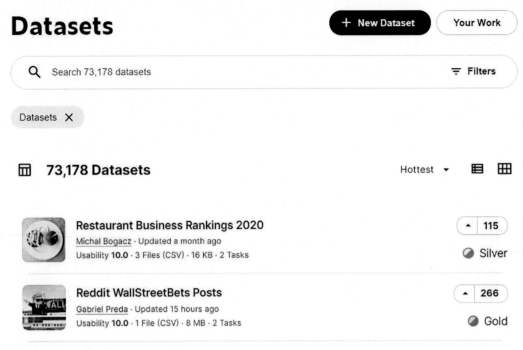

FIG. 2.11   Public datasets on Kaggle.

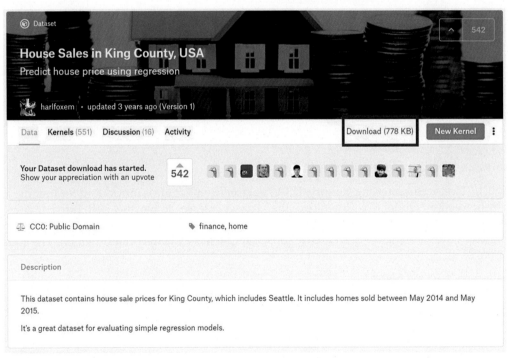

FIG. 2.12   House sales in King County dataset.

- view: an index of how good the view of the property was
- condition: an index on the condition of the house
- grade: an index for rating building construction and design (according to the King County scoring system)
- sqft_above: area of the interior housing space that is above ground level (square feet)
- sqft_basement: area of the interior housing space that is below ground level (square feet)
- yr_built: building time
- yr_renovated: timing of last renovation
- zipcode: ZIP code that the house is in
- lat: latitude coordinates
- long: longitude coordinates
- sqft_living15: square footage of living space recorded in 2015 (implies some renovations)
- sqft_lot15: square footage of land lots recorded in 2015

2.3 Experiment 1: House price prediction

In this section, an FCNN is built and trained on the "House Sales in King County, USA" dataset to predict the price of houses. The network model takes the information of the houses, such as the number of bedrooms, bathrooms, floors, and so on, as the input, and then outputs the price of the house. MSE and Adam are used as the loss function and optimizer of the model, respectively. Fig. 2.13 shows the flowchart of the source code for the house price-prediction model.

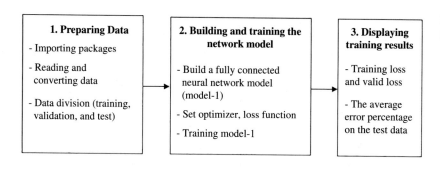

FIG. 2.13 The flowchart of the source code for the house price-prediction model.

2.3.1 Preparing dataset

1. Import necessary packages

```
import os
import numpy as np
import pandas as pd
import tensorflow as tf
import matplotlib.pyplot as plt
from tensorflow import keras
from tensorflow.keras import layers
```

2. Reading and converting data

- Read information

```
data = pd.read_csv(".\dataset\kc_house_data.csv")
# Display the shape of the dataset, a total of 21613 samples, each sample has 21
```

```
kinds of information.
data.shape
```
Result: (21613, 21)

- Display data

```
# Set the number of rows to 25
pd.options.display.max_columns = 25
# display the first five lines (default)
data.head()
```
Result:

	id	date	price	bedrooms	bathrooms	sqft_living	sqft_lot	floors	waterfront	view	condition	grade	sqft_above	sqft_basem
0	7129300520	20141013T000000	221900.0	3	1.00	1180	5650	1.0	0	0	3	7	1180	
1	6414100192	20141209T000000	538000.0	3	2.25	2570	7242	2.0	0	0	3	7	2170	
2	5631500400	20150225T000000	180000.0	2	1.00	770	10000	1.0	0	0	3	6	770	
3	2487200875	20141209T000000	604000.0	4	3.00	1960	5000	1.0	0	0	5	7	1050	
4	1954400510	20150218T000000	510000.0	3	2.00	1680	8080	1.0	0	0	3	8	1680	

- Check the data type

There are five types of data: object (string), boolean, inte ger, float, and categorical.

```
data.dtypes
```
Result:

```
id              int64
date            object
price           float64
bedrooms        int64
bathrooms       float64
sqft_living     int64
sqft_lot        int64
floors          float64
waterfront      int64
view            int64
condition       int64
grade           int64
sqft_above      int64
sqft_basement   int64
yr_built        int64
yr_renovated    int64
zipcode         int64
lat             float64
long            float64
sqft_living15   int64
sqft_lot15      int64
dtype: object
```

▪ Convert data type

Because the date data in the dataset is in a string type and the input of the model only accepts a numeric type, date data including year, month, and day are converted into numeric values through the following code:

```
# convert them to numeric values
data['year'] = pd.to_numeric(data['date'].str.slice(0, 4))
data['month'] = pd.to_numeric(data['date'].str.slice(4, 6))
data['day'] = pd.to_numeric(data['date'].str.slice(6, 8))
#Delete useless data, inplace is to save the updated data to the original place
data.drop(['id'], axis="columns", inplace=True)
data.drop(['date'], axis="columns", inplace=True)
data.head()
```

Result:

aterfront	view	condition	grade	sqft_above	sqft_basement	yr_built	yr_renovated	zipcode	lat	long	sqft_living15	sqft_lot15	year	month	day
0	0	3	7	1180	0	1955	0	98178	47.5112	-122.257	1340	5650	2014	10	13
0	0	3	7	2170	400	1951	1991	98125	47.7210	-122.319	1690	7639	2014	12	9
0	0	3	6	770	0	1933	0	98028	47.7379	-122.233	2720	8062	2015	2	25
0	0	5	7	1050	910	1965	0	98136	47.5208	-122.393	1360	5000	2014	12	9
0	0	3	8	1680	0	1987	0	98074	47.6168	-122.045	1800	7503	2015	2	18

3. Data division

▪ Split data: Divide dataset into three sets: training data, validation data, and test data

```
data_num = data.shape[0]
# Get a random index equal to the number of data,
indexes = np.random.permutation(data_num)
#Randomly divide data into Train, validation and test. The division ratio here is 6:2:2
train_indexes = indexes[:int(data_num *0.6)]
val_indexes = indexes[int(data_num *0.6):int(data_num *0.8)]
test_indexes = indexes[int(data_num *0.8):]
# Retrieve training data, validation data and test data
train_data = data.loc[train_indexes]
val_data = data.loc[val_indexes]
test_data = data.loc[test_indexes]
```

▪ Data normalization

The main function of normalization is to scale different data to the same scale. For example: "The number of bedrooms or bathrooms in the house is about 1 to 5, and the area of the house is about 1500 m^2 to 2500 m^2." Because of the large difference in data scale, it may cause the prediction model to pay more attention to the data with larger values and ignore the data with smaller values. In order to solve this problem, the input data is usually scaled between 0 and 1 or between −1 and 1; this process is called data normalization.

In this experiment, the standard score is used to standardize the data, which is formulated as follows:

$$x_{norm} = \frac{(x - mean)}{std}$$

where, x is a raw score, *mean* is the mean of the population, and *std* is the standard deviation of the population.

The source code for data normalization:

```
train_validation_data = pd.concat([train_data, val_data])
mean = train_validation_data.mean()
std = train_validation_data.std()
train_data = (train_data - mean) / std
val_data = (val_data - mean) / std
```

■ Create the training data in Numpy array format

```
x_train = np.array(train_data.drop('price', axis='columns'))
y_train = np.array(train_data['price'])
x_val = np.array(val_data.drop('price', axis='columns'))
y_val = np.array(val_data['price'])
```

There are a total of 12967 training samples, and each sample has 21 kinds of information.

```
x_train.shape
```

Result: (12967, 21)

2.3.2 Building and training network model

1. Build a FCNN named Model-1

In this example, we construct a network model with three fully connected layers, in which ReLU is used as the activation function in the hidden layers. Since a linear output is required, the output layer does not use any activation function.

```
# Create a fully connected neural network
model = keras.Sequential(name='model-1')
# The first fully connected layer is set to 64 neurons, and the input shape is set to (21, ),
# but in fact the shape of the data we input is (batch_size, 21)
model.add(layers.Dense(64, activation='relu', input_shape=(21,)))
# The second fully connected layer (64 neurons)
```

```
model.add(layers.Dense(64, activation='relu'))
# The output fully connected layer ( 1 neuron).
model.add(layers.Dense(1))
# Display network model structure
model.summary()
```

Result:

```
Model: "model-1"

Layer (type)                 Output Shape              Param #
=================================================================
dense (Dense)                (None, 64)                1408

dense_1 (Dense)              (None, 64)                4160

dense_2 (Dense)              (None, 1)                 65
=================================================================
Total params: 5,633
Trainable params: 5,633
Non-trainable params: 0
```

2. Set the optimizer, loss function, metric function, and callback function.

- Set the optimizer, loss function, metric function

```
model.compile(keras.optimizers.Adam(0.001),
        loss=keras.losses.MeanSquaredError(),
        metrics=[keras.metrics.MeanAbsoluteError()])
```

- Create a directory to save Model

```
model_dir = 'lab2-logs/models/'
os.makedirs(model_dir)  # for creating a folder to save model
```

- Set the callback function:

```
# TensorBoard callback function helps record training information and save as
TensorBoard log file
log_dir = os.path.join('lab2-logs', 'model-1')
model_cbk = keras.callbacks.TensorBoard(log_dir=log_dir)
# ModelCheckpoint helps to save the network model,
model_mckp = keras.callbacks.ModelCheckpoint(model_dir + '/Best-model-1.h5',
                monitor='val_mean_absolute_error',
                save_best_only=True,
                mode='min')
```

3. Training model

```
history = model.fit(x_train, y_train,  # training data
            batch_size=64,  # Batch size is set to 64
            epochs=300,  # Train the entire dataset 300 times
            validation_data=(x_val, y_val),  # Verification information
            callbacks=[model_cbk, model_mckp])
```

Result:

```
Epoch 295/300
12967/12967 [==============================] - 1s 43us/sample - loss: 0.0273 - mean_absolute_error: 0.1221 - val_loss: 0.
1522 - val_mean_absolute_error: 0.2255
Epoch 296/300
12967/12967 [==============================] - 1s 41us/sample - loss: 0.0272 - mean_absolute_error: 0.1215 - val_loss: 0.
1529 - val_mean_absolute_error: 0.2274
Epoch 297/300
12967/12967 [==============================] - 1s 47us/sample - loss: 0.0271 - mean_absolute_error: 0.1213 - val_loss: 0.
1501 - val_mean_absolute_error: 0.2252
Epoch 298/300
12967/12967 [==============================] - 1s 41us/sample - loss: 0.0283 - mean_absolute_error: 0.1237 - val_loss: 0.
1520 - val_mean_absolute_error: 0.2273
Epoch 299/300
12967/12967 [==============================] - 1s 42us/sample - loss: 0.0283 - mean_absolute_error: 0.1220 - val_loss: 0.
1509 - val_mean_absolute_error: 0.2241
Epoch 300/300
12967/12967 [==============================] - 1s 42us/sample - loss: 0.0301 - mean_absolute_error: 0.1258 - val_loss: 0.
1506 - val_mean_absolute_error: 0.2259
```

2.3.3 Displaying training results

1. History

```
history.history.keys()  # View what information is saved in history
```

Result：dict_keys(['loss', 'val_loss', 'val_mean_absolute_error', 'mean_absolute_error'])

2. Draw a line chart of the loss

In "model.compile," the loss function is MSE, so the "loss" and val_loss recorded in the history are the loss values calculated by the MSE.

```
plt.plot(history.history['loss'] , label='train')

plt.plot(history.history['val_loss'] , label='validation')

plt.ylabel('loss')

plt.xlabel('epochs')

plt.legend(loc='upper right')
```

Result:

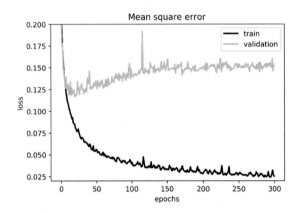

3. Draw a line chart of metrics

In "model.compile," the metric function has been set to MAE, so the network calculates the MAE between the predicted value and the expected output. The mean_absolute_error and val_mean_absolute_error values will be recorded in history.

```
plt.plot(history.history['mean_absolute_error'] , label='train')

plt.plot(history.history['val_mean_absolute_error'] , label='validation')

plt.ylabel('metrics')

plt.xlabel('epochs')

plt.legend(loc='upper right')
```

Result:

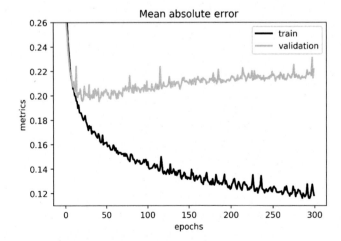

4. The average percentage error on test data

Predict house price on test data and calculate the average percentage error.

```
#Load model
model.load_weights('lab2-logs/models/Best-model-1.h5')
#take out the house price
y_test = np.array(test_data['price'])
# data normalization
test_data = (test_data - mean) / std
# Save the input data in Numpy format
x_test = np.array(test_data.drop('price', axis='columns'))
# Predict on test data
y_pred = model.predict(x_test)
# Convert the prediction results back
y_pred = np.reshape(y_pred * std['price'] + mean['price'], y_test.shape)
# Calculate the mean percentage error
percentage_error = np.mean(np.abs(y_test - y_pred)) / np.mean(y_test) * 100
# Display percentage error
print("Model_1 Percentage Error: {:.2f}%".format(percentage_error))
```

Result: Model_1 Percentage Error: 14.08%

2.4 Introduction to TensorBoard

TensorBoard is a TensorFlow toolkit that provides the visualization and measurement needed for machine learning experimentation such as tracking loss values and the accuracy of the model during training, visualizing the model graph, viewing histograms, and so on. There are two common ways to use TensorBoard: (1) adding "tf.keras.callbacks.TensorBoard" function to create and store logs when training with Model.fit of Keras, and (2) using "tf.summary" API to log information when training with "tf.GradientTape()" or other methods. The graphic interface of the TensorBoard is shown in Fig. 2.14. As shown, there are four main tools in TensorBoard: the Scalars dashboard, Graphs dashboard, Distributions dashboard, and Histograms dashboard.

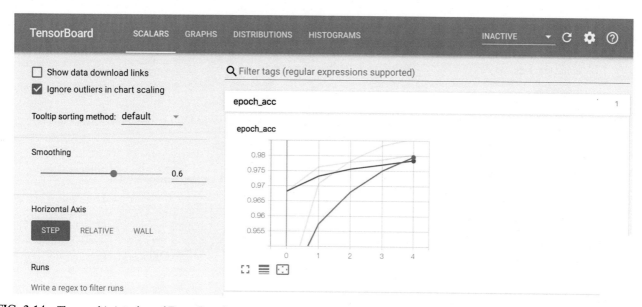

FIG. 2.14 The graphic interface of TensorBoard.

- The Scalars dashboard: helps to track scalar values such as learning rate, loss, accuracy, and so on during training neural networks
- Graphs dashboard: helps to visualize the models built by TensorFlow
- The Distributions and Histograms dashboards: help to display the distribution of the tensor. They are widely used for visualizing weights and biases of the TensorFlow models

The advantages and disadvantages of TensorBoard include:

- Advantages: The information during training the model such as changes in the loss, accuracy, the histograms of weights, biases, and so on, can be tracked and viewed in real time, without having to wait until the training is completed.
- Disadvantages: The information will be written to the log file many times during training the model. If a lot of information is recorded, training time is increased.

When training the house price-prediction model, the TensorBoard callback function, namely, "keras.callbacks. TensorBoard," has been added for creating and storing the log. There are two ways to open the log file. The first way is to directly open the log file on Jupyter Notebook, and the second way is to run TensorBoard through a terminal and then observe results through the browser.

- Open log file with Jupyter Notebook (results are shown in Fig. 2.15)

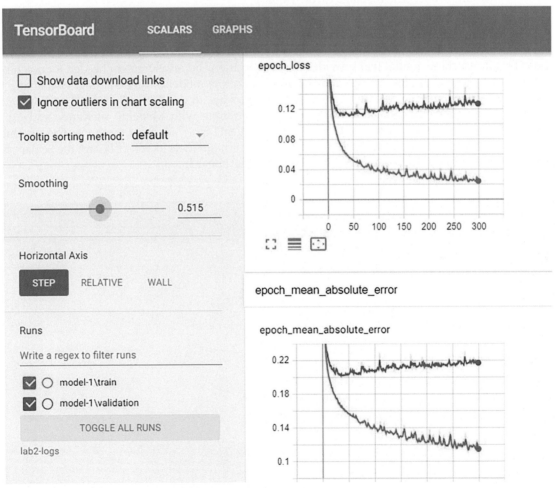

FIG. 2.15   Visualizing metrics (loss and accuracy) on TensorBoard (1).

```
# Loading TensorBoard directly on the jupyter notebook
%load_ext tensorboard
# Run TensorBoard and specify the log file folder as lab2-logs
%tensorboard --logdir lab2-logs
```

Result:

- Open log file with Command line
  - Please go to the location where the TensorBoard log file is stored and run the command below. Note that the result is observed through URL: http://localhost:6006/, as shown in Fig. 2.16.

```
tensorboard --logdir lab2-logs
```

  - The port number can be specified for displaying the result; following the command below, the result can be observed through URL: http://localhost:9527/, as shown in Fig. 2.16.

```
tensorboard --port 9527 --logdir lab2-logs
```

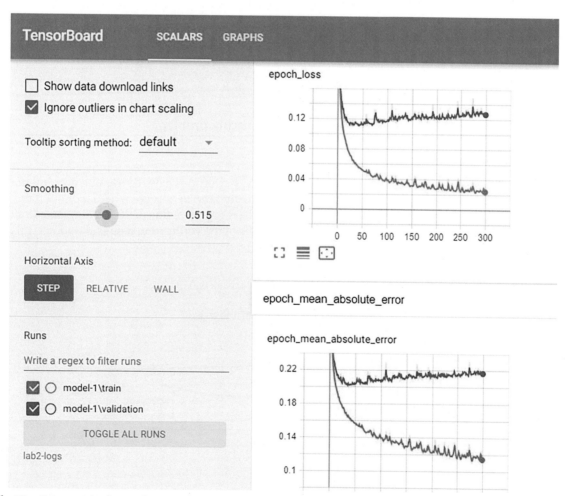

FIG. 2.16   Visualizing metrics (loss and accuracy) on TensorBoard (2).

In addition to metrics such as loss and accuracy, the model graph is also visualized, as shown in Fig. 2.17. We discuss the other visualization functions of TensorBoard such as Images, Text, Audio, and so on in Chapter 7.

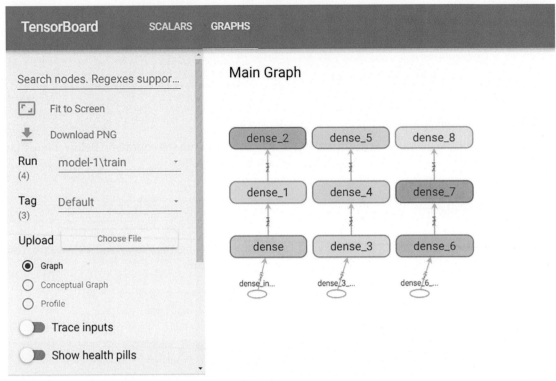

FIG. 2.17   Visualizing the model graph on TensorBoard.

2.5 Experiment 2: Overfitting problem

2.5.1 Introduction to overfitting

Overfitting refers to the network model that obtained very good performance on the training data but that had poor performance on the validation data. The training loss curve is usually used to observe whether or not there is an overfitting problem. Fig. 2.18 presents an overfitting phenomenon, where the loss value of the training data (training error) continues to decrease after a period of training, while the loss value of the verification data (validation error) gradually increases.

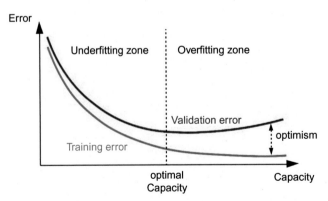

FIG. 2.18   Overfitting phenomenon.

The training result of the house price-prediction model in Section 2.3 is shown in Fig. 2.19, and the overfitting phenomenon can also be observed from the loss curve graph.

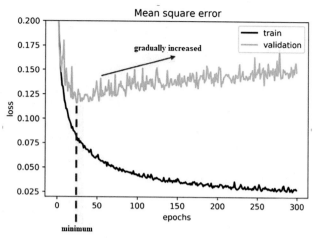

FIG. 2.19 Overfitting problem when training house price-prediction model.

Overfitting usually occurs when the training data is too small in scale or the complexity of the model is too great. As such, adding training data or simplifying the model may improve the problem of overfitting. The three methods to prevent overfitting without increasing the amount of data are:

- Reduce the size of the model: When the number of parameters of the model is reduced, the model with fewer parameters will not be able to easily fit all training data. The model must learn how to use limited parameters to learn an effective feature representation.
- Apply weight regularization: When training a neural network model, the size of the network weights will increase. The longer the network is trained, the larger the network weights will become. The neural network with large weights is usually unstable because even small variation on the inputs can lead to large changes in output [19]. This can be a sign of overfitting training data of the neural network. To solve the overfitting problem, the core idea of weight regularization is to limit the size of the network weights during the training process. To penalize large weights, the first weight size is calculated, and then the calculated result is added to the loss function when training the model. There are two main approaches to calculate weight size: L1 regularization and L2 regularization, also known as weight decay [20].

L1 regularization:

$$\text{Loss}_{Total} = \text{Loss}_{MSE} + \lambda \underbrace{\sum_{j=0}^{M} |w_j|}_{\text{Weight size}}$$

L2 regularization:

$$\text{Loss}_{Total} = \text{Loss}_{MSE} + \lambda \underbrace{\sum_{j=0}^{M} w_j^2}_{\text{Weight size}}$$

where λ is a regularization parameter for controlling the penalty, $\lambda \in [0,1]$, w is the weights of the model, M is the total amount of parameters of the model, and Loss_{MSE} is MSE loss function.

- Apply dropout technique [15]: The dropout technique refers to randomly discarding neurons in the neural network to prevent complex co-adaptions during the training process. When discarding neurons, they are temporarily taken out of the network, and their connections with other neurons are removed as well, as shown in Fig. 2.20. The dropout technique has proved effective in addressing the problem of overfitting and improving the performance of neural networks in many applications such as speech recognition, image classification, and others.

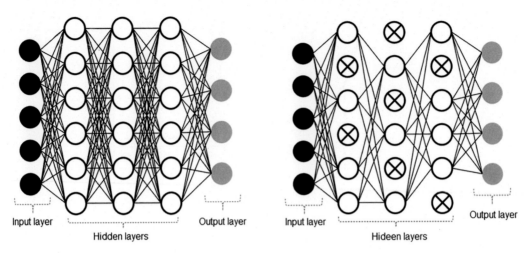

FIG. 2.20 Random dropout.

2.5.2 Code examples

This section continues using the house price-prediction model (Model-1) for testing overfitting. We explore the effect of the three methods for preventing overfitting problems using three modified models based on model-1: (1) model of reducing model size, named Model-2, (2) model of adding weight regularization, named Model-3, and (3) model of adding dropout technique, named Model-4. Table 2.1 shows the architecture of Model-1, Model-2, Model-3, and Model-4.

TABLE 2.1 The architecture of house price-prediction models.

Name	Architecture	Description
Model-1	- Input layer with input shape (,21). - Two hidden layers (fully connected layers); each layer has 64 neurons. - One output layer (fully connected layer) with one neuron.	House price-prediction model in Section 2.3
Model-2	- Input layer with input shape (,21). - Two hidden layers (fully connected layers); each layer has 16 neurons. - One output layer (fully connected layer) with one neuron.	Model of reducing the Model-1 size
Model-3	- Input layer with input shape (,21). - Two hidden layers (fully connected layers); each layer has 64 neurons and L2 regularization. - One output layer (fully connected layer) with one neuron.	Model of adding weights' regularization
Model-4	- Input layer with input shape (,21). - Two hidden layers (fully connected layers); each layer has 64 neurons; randomly discard 30% of neurons in each layer. - One output layer (fully connected layer) with one neuron.	Model of adding dropout technique

1. Model-2: model of reducing the Model-1 size

```python
# Create model-2
model_2 = keras.Sequential(name='model-2')
# first hidden fully connected layer with 16 neurons
model_2.add(layers.Dense(16, activation='relu', input_shape=(21,)))
# second hidden fully connected layer with 16 neurons
model_2.add(layers.Dense(16, activation='relu'))
# output fully connected layer with 1 neurons
model_2.add(layers.Dense(1))


# Set the optimizer, loss function and metrics function for training
model_2.compile(keras.optimizers.Adam(0.001),
        loss=keras.losses.MeanSquaredError(),
        metrics=[keras.metrics.MeanAbsoluteError()])


# Set callback function
log_dir = os.path.join('lab2-logs', 'model-2')
model_cbk = keras.callbacks.TensorBoard(log_dir=log_dir)
model_mckp = keras.callbacks.ModelCheckpoint(model_dir + '/Best-model-2.h5',
                monitor='val_mean_absolute_error',
                save_best_only=True,
                mode='min')
# Train model-2
model_2.fit(x_train, y_train,
        batch_size=64,
        epochs=300,
        validation_data=(x_val, y_val),
        callbacks=[model_cbk, model_mckp])
```

```
Epoch 296/300
12967/12967 [==============================] - 1s 41us/sample - loss: 0.0725 - mean_absolute_error: 0.1754 - val_loss: 0.
1202 - val_mean_absolute_error: 0.2006
Epoch 297/300
12967/12967 [==============================] - 1s 43us/sample - loss: 0.0720 - mean_absolute_error: 0.1760 - val_loss: 0.
1188 - val_mean_absolute_error: 0.2023
Epoch 298/300
12967/12967 [==============================] - 1s 43us/sample - loss: 0.0729 - mean_absolute_error: 0.1765 - val_loss: 0.
1210 - val_mean_absolute_error: 0.2018
Epoch 299/300
12967/12967 [==============================] - 1s 43us/sample - loss: 0.0721 - mean_absolute_error: 0.1758 - val_loss: 0.
1258 - val_mean_absolute_error: 0.2015
Epoch 300/300
12967/12967 [==============================] - 1s 48us/sample - loss: 0.0737 - mean_absolute_error: 0.1769 - val_loss: 0.
1200 - val_mean_absolute_error: 0.2032

<tensorflow.python.keras.callbacks.History at 0x7ff73c7b4940>
```

2. Model-3: model of adding weight regularization

```python
# Create a network model
model_3 = keras.Sequential(name='model-3')
# first hidden fully connected layer with 64 neurons, adding L2 regularization
model_3.add(layers.Dense(64, kernel_regularizer=keras.regularizers.l2(0.001),
                activation='relu', input_shape=(21,)))
# Second hidden fully connected layer with 64 neurons, adding L2 regularization
model_3.add(layers.Dense(64, kernel_regularizer=keras.regularizers.l2(0.001),
                activation='relu'))
# hidden output fully connected layer with 1 neuron
model_3.add(layers.Dense(1))


# Set the optimizer, loss function and metric function for training
model_3.compile(keras.optimizers.Adam(0.001),
        loss=keras.losses.MeanSquaredError(),
        metrics=[keras.metrics.MeanAbsoluteError()])


# Set callback function
log_dir = os.path.join('lab2-logs', 'model-3')
model_cbk = keras.callbacks.TensorBoard(log_dir=log_dir)
model_mckp = keras.callbacks.ModelCheckpoint(model_dir + '/Best-model-3.h5',
                        monitor='val_mean_absolute_error',
                        save_best_only=True,
                        mode='min')


# Train model-3
model_3.fit(x_train, y_train,
        batch_size=64,
        epochs=300,
        validation_data=(x_val, y_val),
        callbacks=[model_cbk, model_mckp])
```

```
Epoch 296/300
12967/12967 [==============================] - 1s 44us/sample - loss: 0.0687 - mean_absolute_error: 0.1528 - val_loss: 0.
1293 - val_mean_absolute_error: 0.1924
Epoch 297/300
12967/12967 [==============================] - 1s 45us/sample - loss: 0.0744 - mean_absolute_error: 0.1571 - val_loss: 0.
1323 - val_mean_absolute_error: 0.1927
Epoch 298/300
12967/12967 [==============================] - 1s 45us/sample - loss: 0.0756 - mean_absolute_error: 0.1587 - val_loss: 0.
1440 - val_mean_absolute_error: 0.1964
Epoch 299/300
12967/12967 [==============================] - 1s 44us/sample - loss: 0.0717 - mean_absolute_error: 0.1555 - val_loss: 0.
1275 - val_mean_absolute_error: 0.1890
Epoch 300/300
12967/12967 [==============================] - 1s 44us/sample - loss: 0.0691 - mean_absolute_error: 0.1528 - val_loss: 0.
1320 - val_mean_absolute_error: 0.1907

<tensorflow.python.keras.callbacks.History at 0x7ff73c18fc18>
```

3. Model-4: model of adding dropout

```python
# Create model-4
model_4 = keras.Sequential(name='model-4')
# first hidden fully connected layer with 64 neurons,
model_4.add(layers.Dense(64, activation='relu', input_shape=(21,)))
#randomly discard 30% neurons
model_4.add(layers.Dropout(0.3))
# second hidden fully connected layer with 64 neurons
model_4.add(layers.Dense(64, activation='relu'))
# randomly discard 30% neurons
model_4.add(layers.Dropout(0.3))
#Output fully connected layer with 1 neurons
model_4.add(layers.Dense(1))


# Set the optimizer, loss function and indicator function for training
model_4.compile(keras.optimizers.Adam(0.001),
        loss=keras.losses.MeanSquaredError(),
        metrics=[keras.metrics.MeanAbsoluteError()])


# Set callback function
log_dir = os.path.join('lab2-logs', 'model-4')
model_cbk = keras.callbacks.TensorBoard(log_dir=log_dir)
model_mckp = keras.callbacks.ModelCheckpoint(model_dir + '/Best-model-4.h5',
                    monitor='val_mean_absolute_error',
                    save_best_only=True,
                    mode='min')


# Train model-4
model_4.fit(x_train, y_train,
        batch_size=64,
        epochs=300,
```

```python
        validation_data=(x_val, y_val),
        callbacks=[model_cbk, model_mckp])
```

After training, the trained Model-2, Model-3, and Model-4 are verified on the test data.

    1. Model-2:

```
model_2.load_weights('lab2-logs/models/Best-model-2.h5')
y_pred = model_2.predict(x_test)
y_pred = np.reshape(y_pred * std['price'] + mean['price'], y_test.shape)
percentage_error = np.mean(np.abs(y_test - y_pred)) / np.mean(y_test) * 100
print("Model_2 Percentage Error: {:.2f}%".format(percentage_error))
```

Result: Model_2 Percentage Error: 13.15%

    2. Model-3:

```
model_3.load_weights('lab2-logs/models/Best-model-3.h5')
y_pred = model_3.predict(x_test)
y_pred = np.reshape(y_pred * std['price'] + mean['price'], y_test.shape)
percentage_error = np.mean(np.abs(y_test - y_pred)) / np.mean(y_test) * 100
print("Model_3 Percentage Error: {:.2f}%".format(percentage_error))
```

Result: Model_3 Percentage Error: 12.89%

    3. Model-4:

```
model_4.load_weights('lab2-logs/models/Best-model-4.h5')
y_pred = model_4.predict(x_test)
y_pred = np.reshape(y_pred * std['price'] + mean['price'], y_test.shape)
percentage_error = np.mean(np.abs(y_test - y_pred)) / np.mean(y_test) * 100
print("Model_4 Percentage Error: {:.2f}%".format(percentage_error))
```

Result: Model_4 Percentage Error: 13.33%

2.5.3 Visualization with TensorBoard

In this section, we use TensorBoard to observe and analyze the training results of Model-1, Model-2, Model-3, and Model-4 in Section 2.5.2.

- Uncheck all the training data records in the lower-left corner of TensorBoard, displaying only the validation data records, as shown in Fig. 2.21. In Fig. 2.21, model-1/validation with the orange line chart represents an overfitting model; model-2/validation with the maroon line chart represents the model of reducing parameters; model-3/validation with the cyan line chart represents the model of adding L2 regularization; and model-4/validation with the green line chart represents the model of using the dropout technique.
- Adjust the smoothing ratio to zero to display the most original data without modification; note that setting the smoothing to zero is more convenient for finding the lowest point, as shown in Fig. 2.22.

The experimental results prove that three methods are capable of solving the overfitting problem of the original model and improving the performance of the model. Among them, Model-4 applying the dropout technique has the lowest loss value, while Model-3 using L2 normalization has the lowest percentage error.

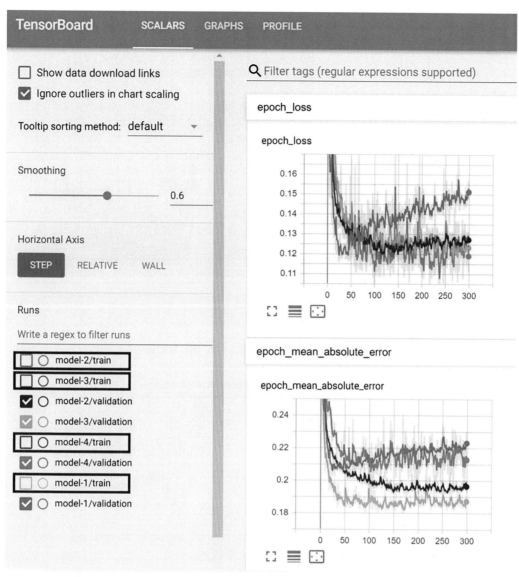

FIG. 2.21   TensorBoard Scalars (1).

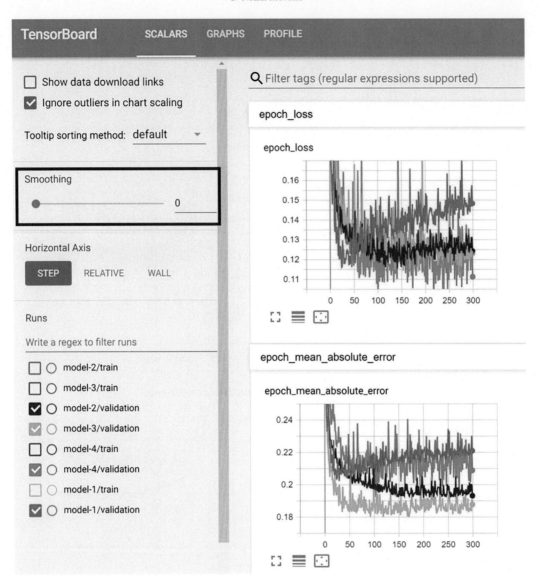

FIG. 2.22   TensorBoard Scalars (2).

References

[1] W.S. McCulloch, W. Pitts, A logical calculus of the ideas immanent in nervous activity, Bull. Math. Biophys. 5 (4) (1943) 115–133.

[2] N. Rochester, J. Holland, L. Haibt, W. Duda, Tests on a cell assembly theory of the action of the brain, using a large digital computer, IRE Trans. Inform. Theory 2 (3) (1956) 80–93.

[3] B.W.A.C. Farley, W. Clark, Simulation of self-organizing systems by digital computer, Trans. IRE Prof. Group Inform. Theory 4 (4) (1954) 76–84.

[4] F. Rosenblatt, The perceptron: a probabilistic model for information storage and organization in the brain, Psychol. Rev. 65 (6) (1958) 386–408.

[5] J.J. Weng, N. Ahuja, T.S. Huang, Learning recognition and segmentation of 3-D objects from 2-D images, in: (4th) International Conference on Computer Vision, IEEE, 1993, pp. 121–128.

[6] T. Joachims, Text categorization with support vector machines: learning with many relevant features, in: European Conference on Machine Learning, Springer, Berlin, Heidelberg, 1998, pp. 137–142.

[7] C. Cortes, V. Vapnik, Support-vector networks, Mach. Learn. 20 (3) (1995) 273–297.

[8] P. Werbos, Beyond Regression: New Tools for Prediction and Analysis in the Behavioral Sciences, PhD thesis Harvard University, Cambridge, MA, 1974.

[9] A. Krizhevsky, I. Sutskever, G.E. Hinton, Imagenet classification with deep convolutional neural networks, in: Advances in Neural Information Processing Systems, 2012, pp. 1097–1105.

[10] J. Deng, W. Dong, R. Socher, L.-J. Li, K. Li, L. Fei-Fei, Imagenet: a large-scale hierarchical image database, in: Proceedings of the IEEE Conference on Computer Vision and Pattern Recognition, 2009, pp. 248–255.

[11] X. Glorot, A. Bordes, Y. Bengio, Deep sparse rectifier neural networks, in: International Conference on Artificial Intelligence and Statistics, 2011, pp. 315–323.

[12] K. He, X. Zhang, S. Ren, J. Sun, Delving deep into rectifiers: Surpassing human-level performance on imagenet classification, in: Proceedings of the IEEE International Conference on Computer Vision, 2015, pp. 1026–1034.

[13] B. Recht, R. Roelofs, L. Schmidt, V. Shankar, Do cifar-10 classifiers generalize to cifar-10? arXiv preprint arXiv:1806.00451, (2018).

[14] J. Janke, M. Castelli, A. Popovic, Analysis of the proficiency of fully connected neural networks in the process of classifying digital images. benchmark of different classification algorithms on high-level image features from convolutional layers. Expert Syst. Appl. 135 (2019) 12–38, [Online]. Available https://doi.org/10.1016/j.eswa.2019.05.058.

[15] N. Srivastava, G.E. Hinton, A. Krizhevsky, I. Sutskever, R. Salakhustdinov, Dropout: a simple way to prevent neural networks from overfitting, J. Mach. Learn. Res. 15 (1) (2014) 1929–1958.

[16] I. Sutskever, J. Martens, G. Dahl, G. Hinton, On the importance of initialization and momentum in deep learning, in: International Conference on Machine Learning, 2013, pp. 1139–1147.

[17] J. Duchi, E. Hazan, Y. Singer, Adaptive subgradient methods for online learning and stochastic optimization, J. Mach. Learn. Res. (2011) 2121–2159.

[18] D. Kinga, J.B. Adam, A method for stochastic optimization, in: International Conference on Learning Representations (ICLR), vol. 5, 2015.

[19] R. Reed, R.J. Marks II, Neural Smithing: Supervised Learning in Feedforward Artificial Neural Networks, MIT Press, 1999, 269.

[20] A. Krogh, J.A. Hertz, A simple weight decay can improve generalization, in: Advances in Neural Information Processing Systems, 1992, pp. 950–957.

3

Binary classification problem

3.1 Machine learning algorithms

Machine learning is a subfield of computer science divided into four groups according to learning style: supervised learning, unsupervised learning, semi-supervised learning, and reinforcement learning. The following introduces the methods and applications of each type.

- Supervised learning: An algorithm teaches the network model by using a given labeled dataset that consists of samples and corresponding labels. This algorithm is categorized into two main types, including classification and regression, and is widely applied in numerous computer vision applications such as object detection, image segmentation, speech recognition, and so on. The most commonly used supervised learning methods are:
  - k-nearest neighbors algorithm (KNN) [1–4]
  - decision trees [5–8] and random forests [9–11]
  - support vector machines (SVMs) [12–15]
  - deep neural networks (DNNs) [16–19]
- Unsupervised learning: Unlike the supervised learning algorithm, the unsupervised learning algorithm only uses the dataset without labels or annotations for drawing inference. The common tasks of the unsupervised learning algorithm include clustering and dimension reduction. Clustering is used to classify the input data into subgroups based on the correlation between the samples in each group. Dimension reduction is the process of transforming data from a high-dimensional space into a low-dimensional space while maintaining the structure and characteristics of the data. The following are commonly used unsupervised learning methods:
  - For clustering problem
    - k-means algorithm [20–23]
    - hierarchical cluster algorithm [24–26]
  - For dimension reduction problem
    - principal component analysis (PCA) algorithm [27–29]
    - kernel PCA [30–32]
- Semi-supervised learning: An approach that uses a set of data containing a larger amount of unlabeled data and a small amount of labeled data for training the models. Typically, semi-supervised learning algorithms combine supervised learning and unsupervised learning in an attempt to improve performance in one of these two

techniques through the use of information related to the other [33]. These algorithms are commonly applied for classification problems in which labeled data is scarce or difficult to achieve, such as computer-aided diagnosis, grammatical tagging, and so on.

- Reinforcement Learning: This technique has attracted much attention, as the Google DeepMind team has successfully applied reinforcement learning to Atari games, AlphaGo [34,35], and robots. Reinforcement learning is a technique that allows an agent to learn and take actions in an interactive environment to maximize the expected cumulative reward. Specifically, at time t, the state S_t and reward R_t from the environment are sent to the agent, then the agent makes its own decision to select an action A_t from a set of available actions to send back to the environment. The process is repeated with a new state S_{t+1} and reward R_{t+1} of the environment, as shown in Fig. 3.1.

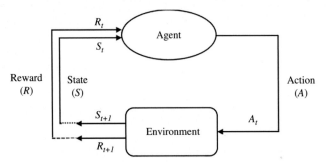

FIG. 3.1   Reinforcement learning.

3.2 Binary classification problem

3.2.1 Introduction to binary classification

Binary classification is one of the most common problems in machine learning. In its basic form, it refers to the model of predicting an input sample into one of two classes. The binary classification models are often utilized to infer the probability of an event or a certain class such as disease or not, male or female, win or lose, pass or fail, and so on. Fig. 3.2 shows an example of binary classification, where the "cat" image is passed through a network model for computing and inferring the probability p. Based on the predicted p, the model determines whether the input sample belongs to the "Cat" class or "Not Cat" class. Suppose that $p=0.9$; this means there is 90% confidence that the input sample belongs to the "Cat" class, and 10% confidence that the input sample belongs to the "Not Cat" class.

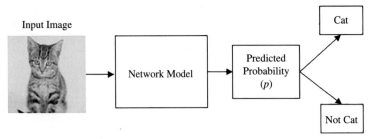

FIG. 3.2   Example of binary classification.

3.2.2 Binary classification model

The binary classification models predict an input sample into one of two classes by modeling the probability of the sample; therefore, its output is only a single probability score. Fig. 3.3. presents an overview of the binary classification model.

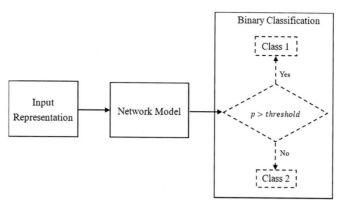

FIG. 3.3 Overview of binary classification model; p is the predicted probability of the network model.

1. Input representation

As mentioned in Section 3.1, the classification problem belongs to the supervised learning method, which requires input pairs of samples and corresponding labels for learning. However, most of machine learning algorithms, especially neural networks, cannot directly work with labeled or categorical data; they need to be converted into numbers. Integers can be used to encode this data, which may work for the tasks where the categories have a natural ordinal relationship with each other, such as the labels for the density of fog: "light," "medium," and "heavy." If there is no a natural ordinal relationship between the categories, such as the labels "table" and "person," this may be resulting in poor performance of the model when using an integer-encoding technique because of allowing the model to assume a natural ordering between categories.

One-hot encoding is a method of representing categorical data by using a group of bits containing bits 0 and 1. If there are N classes, N groups of bits will be built, in which each group is represented by N-1 bits 0 and only one bit 1. This helps the representation of each category to be completely independent and more expressive. For example, given eight classes including people, dog, cat, bird, flower, aircraft, car, and truck, as listed in the first column of Table 3.1. If the corresponding numbers 1, 2, 3, 4, 5, 6, 7, and 8 are used to represent these classes, as shown in the second column of the Table 3.1, the "dog" class will be closer to the "people" class and the "cat" class than the other classes. This is because 2 is closer to 1 and 3 than the other numbers in the relationship. However, this is not reasonable because there is not a natural, original relationship between these eight classes. By using one-hot encoding to represent these eight classes, each class is completed independently, as shown in the third column of Table 3.1. In the next section, both one-hot encoding and integer-encoding techniques are applied to represent the categorical data of the Pokémon combat prediction model for exploring the effect of each technique.

TABLE 3.1 Numerical representation and one-hot encoding representation.

Class	Integer encoding	One-hot encoding
People	1	00000001
Dog	2	00000010
Cat	3	00000100
Bird	4	00001000
Flower	5	00010000
Aircraft	6	00100000
Car	7	01000000
Truck	8	10000000

2. Network Model

In the binary classification problem, the network model is responsible for learning features from the input samples and outputting a single probability for predicting one of two classes. In the following, we introduce the network architecture and loss function for training binary classification models.

(a) Network architecture

Because of strong learning capacity, neural networks have been applied to the binary classification problem to achieve high performance [36]. As introduced in Chapter 2, a neural network is composed of an input layer, hidden layers, and an output layer, where each layer contains neurons for connection and computation. For binary classification models, depending on each problem and the characteristics of the data, the hyperparameters for building the neural network (e.g., the number of neurons in the input layer, the number of neurons in each hidden layer, the numbers of hidden layers, etc.) may be different, but the number of neurons in the output layer is only one, as shown in Fig. 3.4. Chapter 7 discusses selecting an optimal neural network architecture for a specific problem like classification. In Section 3.3 of this chapter, a neural network with five layers is introduced for predicting the result (the winner, the loser) of Pokémon combats.

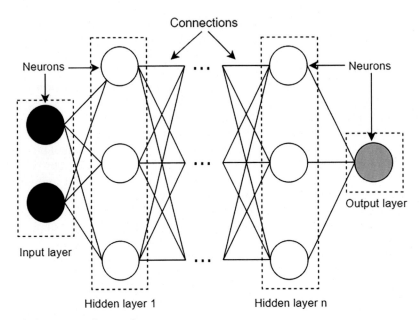

FIG. 3.4   The architecture of a binary neural network.

(b) Loss function

Cross-entropy (CE) and mean squared error (MSE) are widely used loss functions for training machine learning models. While MSE is usually employed for linear regression models (see Chapter 2), CE is a priority option for classification models.

The formulas for MSE and CE are:

$$MSE = \frac{\sum_{i=1}^{N}(y_i - \hat{y}_i)^2}{N}$$

$$CE = -\frac{\sum_{i=1}^{N}\sum_{j=0}^{C}y_{i,j}\log\hat{y}_{i,j}}{N},$$

where y is the expected output, $\hat{y}$ is the prediction output of the model, C is the number of categories, and N is the amount of data in a batch.

Fig. 3.5 shows a comparison of CE and MSE. As shown, if it is a correct prediction ($\hat{y} = 1$), both the loss values of MSE and CE will be 0; if it is a wrong prediction ($\hat{y} = 0$), the maximum loss value of MSE is 1, while that of CE is infinity. Since the classification refers to the task of predicting one of two or more classes while linear regression refers to the task of predicting continuous values, the decision boundary between two classes in the classification problem is larger than that of the linear regression problem. If MSE is applied for classification, it cannot penalize misclassifications enough, leading to low efficiency in training the network models.

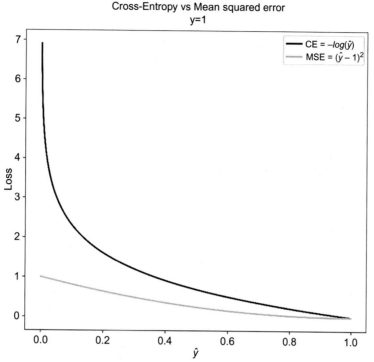

FIG. 3.5 Difference between cross-entropy (CE) and mean squared error (MSE).

The classification problem can be divided into two categories: binary classification problem and multi-class classification problem. Binary cross-entropy (BCE) is usually used as a loss function for the binary classification models, and categorical cross-entropy (CCE) is often employed as the loss function for the multi-class classification models. This chapter focuses on a binary classification problem, so BCE is introduced and used as a loss function for training the model. Chapter 4 further discusses CCE.

The BCE is a combination of sigmoid activation and CE loss, also called sigmoid cross-entropy loss, as shown in Fig. 3.6.

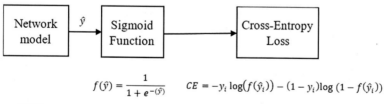

FIG. 3.6 Binary cross-entropy (BCE).

The formula for BCE is:

$$BCE = -\frac{\sum_{i=1}^{N} \left[y_i \log\left(f\left(\hat{y}_i\right)\right) + (1 - y_i) \log\left(1 - f\left(\hat{y}_i\right)\right) \right]}{N},$$

where, y is expected output, $\hat{y}$ is the predicted value of the network model, N is the amount of data in a batch, and f is sigmoid function.

When training machine learning models using BCE as an objective function, the returned loss values are between 0 and 1, and it is better to get a low loss value if the model predicts the label correctly. As can be observed in Fig. 3.7, in both cases of the expected output, y is equal to 1 or 0. The closer the predicted value ($\hat{y}$) is to the expected output (y), the smaller the loss value becomes.

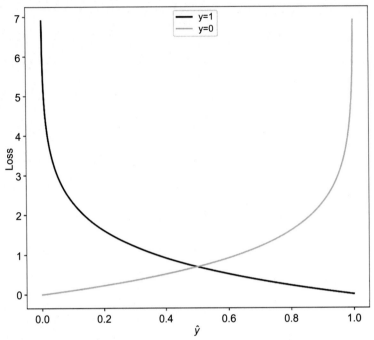

FIG. 3.7   The relationship between loss value and predicted output ($\hat{y}$) of the model using binary cross-entropy (BCE).

3. Binary classification

The output of binary classification models is single probability score. The model determines whether the input sample belongs to the first class or the second class by comparing its predicted probability (p) with a threshold, as shown in Fig. 3.8. Normally, the threshold is set to a value of 0.5 as the default for the predicted probabilities in the range between 0 and 1. If the predicted probability is greater than the threshold, the input sample is classified into the first class; otherwise, it is assigned to the second class.

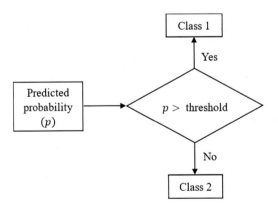

FIG. 3.8   Binary classification.

3.3 Experiment: Pokémon combat prediction

In this section, a Pokémon combat prediction model is implemented to illustrate a binary classification model. In a combat, a winner can be predicted based on the characteristics of Pokémon such as the amount of blood, attack force, defense points, and so on. Therefore, the Pokémon characteristics are used as the inputs of the Pokémon combat prediction model. The BCE loss function introduced earlier is employed for training the model. The model predicts the winner of the combat between the first Pokémon and the second Pokémon based on output probability value. If the predicted probability is less than 0.5, the winner is the first Pokémon. On the contrary, the winner is the second Pokémon.

In order to verify the effect of the one-hot encoding technique, we implement two scenarios of the Pokémon combat prediction model:

(1) In the first scenario, the integer-encoding technique is used to represent the attributes of the Pokémon for training the model, named Model-1.
(2) In the second scenario, the one-hot-encoding technique is employed to represent the attributes of Pokémon for the training model, named Model-2.

3.3.1 Introduction to Pokémon-Weedle's cave dataset

The example in this chapter uses a Kaggle dataset, namely, the "Pokémon-Weedle's Cave" dataset for training and evaluating the Pokémon combat prediction model. Please access https://www.kaggle.com/terminus7/pokemon-challenge and click "Download" to download the dataset, as shown in Fig. 3.9.

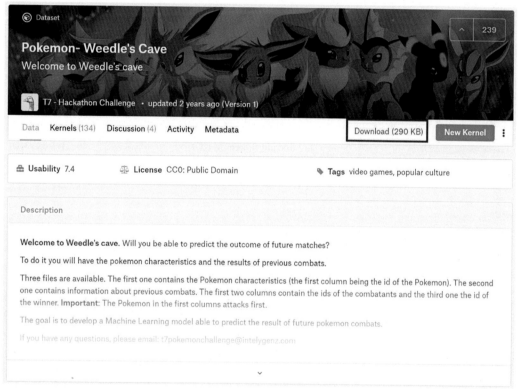

FIG. 3.9 Pokémon-Weedle's Cave dataset page.

To predict the outcome of future matches between two Pokémon, the Pokémon-Weedle's Cave dataset provides the Pokémon characteristics and the results of previous combats for training models.

1. Pokémon data

Pokémon-Weedle's Cave dataset has 800 Pokémon. Table 3.2 displays the information of the first five Pokémon. The characteristics of each Pokémon are as follows:
- Name: Pokémon name
- Type 1: The first attribute
- Type 2: The second attribute

- HP: Hitpoint
- Attack: Attack force
- Defense: Defense point
- Sp. Atk: Special attack force
- Sp. Def: Special defense point
- Speed: Speed of Pokémon
- Generation: Evolutionary stage
- Legendary: Legendary Pokémon

TABLE 3.2   Characteristics of the first five Pokémon.

ID	Name	Type 1	Type 2	HP	Attack	Defense	Sp atk	Sp def	Speed	Generation	Legendary
1	Bulbasaur	Grass	Poison	45	49	49	65	65	45	1	False
2	Ivysaur	Grass	Poison	60	62	63	80	80	60	1	False
3	Venusaur	Grass	Poison	80	82	83	100	100	80	1	False
4	MegaVenusaur	Grass	Poison	80	100	123	122	120	80	1	False
5	Charmander	Fire		39	52	43	60	50	65	1	False

2. Combat data

Pokémon-Weedle's Cave dataset provides the outcome of 50,000 Pokémon combats. Table 3.3 lists the information of the first five combats, where the numbers represent the ID of each Pokémon.

TABLE 3.3   The first five combats in Pokémon-Weedle's Cave dataset.

First Pokemon (ID)	Second Pokemon (ID)	Winner (ID)
266	298	298
702	701	701
191	668	668
237	683	683
151	231	151

The detail information of each Pokémon can be found from the ID of the Pokémon in combat, as shown in Table 3.4. As shown, the first Pokémon with ID of 5 corresponds to Charmander Pokémon; the second Pokémon with ID of 1 corresponds to Bulbasaur Pokémon; and the winner of the combat is Charmander Pokémon.

TABLE 3.4   Combat data and Pokémon characteristics.

(a) Combat data

First Pokemon(ID)	Second Pokemon(ID)	Winner(ID)
5	1	5

(b) Pokemon characteristics

ID	Name	Type1	Type2	HP	Attack	Defense	Sp Atk	Sp Def	Speed	Generation	Legendary
1	Bulbasaur	Grass	Poison	45	49	49	65	65	45	1	False
2	Ivysaur	Grass	Poison	60	62	63	80	80	60	1	False
3	Venusaur	Grass	Poison	80	82	83	100	100	80	1	False
4	MegaVenusaur	Grass	Poison	80	100	123	122	120	80	1	False
Winner 5	Charmander	Fire		39	52	43	60	50	65	1	False

The training dataset contains training samples and corresponding ground-truth labels as follows:

- Training samples: First Pokémon (Type 1, Type 2, HP, Attack, Defense, Sp. Atk, Sp. Def, Speed, Generation, Legendary) and second Pokémon (Type 1, Type 2, HP, Attack, Defense, Sp. Atk, Sp. Def, Speed, Generation, Legendary).
- Ground-truth labels (or expected output): First Pokémon or second Pokémon; 0 means the first Pokémon wins, 1 means the second Pokémon wins.

Supplementary explanation

There are 18 attributes (Type1 and Type2) of Pokémon in Pokémon-Weedle's Cave dataset including glass, fire, water, bug, normal, poison, electric, ground, fairy, fighting, psychic, rock, ghost, ice, dragon, steel, dark, and flying. The numbers from 0 to 17 can be used to represent these attributes. However, there is no a natural ordinal relationship between attributes (Section 3.2.2), so it would be better to use the one-hot encoding technique to encode these attributes.

3.3.2 Code examples

Fig. 3.10 shows the flowchart of the source code for the Pokémon combat prediction model.

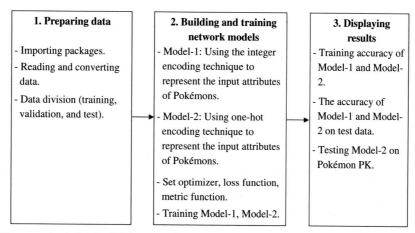

FIG. 3.10 Flowchart of the source code for the Pokémon combat prediction model.

1. Preparing data
   (a) Import packages

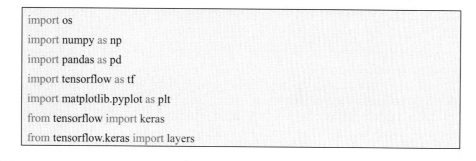

```
import os
import numpy as np
import pandas as pd
import tensorflow as tf
import matplotlib.pyplot as plt
from tensorflow import keras
from tensorflow.keras import layers
```

(b) Reading and converting data
- Read Pokémon data from CSV file

```
pokemon_df = pd.read_csv('./dataset/pokemon-challenge/pokemon.csv')
pokemon_df.head()
```

Result:

	#	Name	Type 1	Type 2	HP	Attack	Defense	Sp. Atk	Sp. Def	Speed	Generation	Legendary
0	1	Bulbasaur	Grass	Poison	45	49	49	65	65	45	1	False
1	2	Ivysaur	Grass	Poison	60	62	63	80	80	60	1	False
2	3	Venusaur	Grass	Poison	80	82	83	100	100	80	1	False
3	4	Mega Venusaur	Grass	Poison	80	100	123	122	120	80	1	False
4	5	Charmander	Fire	NaN	39	52	43	60	50	65	1	False

- Set "#" as index value

```
pokemon_df= pokemon_df.set_index("#")
pokemon_df.head()
```

Result:

#	Name	Type 1	Type 2	HP	Attack	Defense	Sp. Atk	Sp. Def	Speed	Generation	Legendary
1	Bulbasaur	Grass	Poison	45	49	49	65	65	45	1	False
2	Ivysaur	Grass	Poison	60	62	63	80	80	60	1	False
3	Venusaur	Grass	Poison	80	82	83	100	100	80	1	False
4	Mega Venusaur	Grass	Poison	80	100	123	122	120	80	1	False
5	Charmander	Fire	NaN	39	52	43	60	50	65	1	False

- Read combats data

```
combats_df = pd.read_csv('./dataset/pokemon-challenge/combats.csv')
combats_df.head()
```

Result:

	First_pokemon	Second_pokemon	Winner
0	266	298	298
1	702	701	701
2	191	668	668
3	237	683	683
4	151	231	151

- Check if there is any missing data in Pokémon data
  - Pokémon data contains 800 Pokémon samples, but "Name" and "Type 2" of Pokémon in the Pokémon data have missing information.
  - "Name" information may be missing from the original dataset, but it does not affect training, because the name will not be used during training.
  - Type 2: the second attribute of Pokémon is missing information because some Pokémon do not have this attribute, so the data needs to be filled into the missing field.

```
pokemon_df.info()
```

Result:

```
<class 'pandas.core.frame.DataFrame'>
Int64Index: 800 entries, 1 to 800
Data columns (total 11 columns):
Name          799 non-null object      missing 1 sample
Type 1        800 non-null object
Type 2        414 non-null object      missing 386 samples
HP            800 non-null int64
Attack        800 non-null int64
Defense       800 non-null int64
Sp. Atk       800 non-null int64
Sp. Def       800 non-null int64
Speed         800 non-null int64
Generation    800 non-null int64
Legendary     800 non-null bool
dtypes: bool(1), int64(7), object(3)
memory usage: 69.5+ KB
```

- View the number of Pokémon without second attribute (Type 2)

  By setting the parameter "dropna = False," the missing data (NaN) can be taken into account. NaN means that Pokémon does not have the second attribute.

```
pokemon_df["Type 2"].value_counts(dropna=False)
```

Result:

```
NaN          386          there are 386 Pokemon without the second attribute
Flying        97
Ground        35
Poison        34
Psychic       33
Fighting      26
Grass         25
Fairy         23
Steel         22
Dark          20
Dragon        18
Water         14
Rock          14
Ice           14
Ghost         14
Fire          12
Electric       6
Normal         4
Bug            3
Name: Type 2, dtype: int64
```

- Fill in missing data: replacing "NaN" with "empty"

```
pokemon_df["Type 2"].fillna('empty',inplace=True)
pokemon_df["Type 2"].value_counts()
```

Result:

empty	386
Flying	97
Ground	35
Poison	34
Psychic	33
Fighting	26
Grass	25
Fairy	23
Steel	22
Dark	20
Dragon	18
Water	14
Rock	14
Ice	14
Ghost	14
Fire	12
Electric	6
Normal	4
Bug	3

Name: Type 2, dtype: int64

NaN → empty

- Checking data type

Type1, Type2, and Legendary are input data of the network model. Because of the difference in data types, they cannot be directly inputted into the model. Data conversion is required.

Print the data type of input data:

```
print(combats_df.dtypes)
print('-' * 30)
print(pokemon_df.dtypes)
```

Result:

First_pokemon	int64
Second_pokemo	int64
Winner	int64

dtype: object

Name	objec
Type 1	objec
Type 2	objec
HP	int64
Attack	int64
Defense	int64
Sp. Atk	int64
Sp. Def	int64
Speed	int64
Generation	int64
Legendary	bool

dtype: object

- Conversion of data type
  - Type1, Type2: Convert data type of "Type1" and "Type2" from "object" to "**category**"
  - Legendary: Convert data type of "Legendary" from "**bool**" to "**int**." The data representation will be changed from "False" and "True" to 0 and 1, respectively

```python
# Convert Type1 to category type
pokemon_df['Type 1'] = pokemon_df['Type 1'].astype('category')
# Convert Type2 to category type
pokemon_df['Type 2'] = pokemon_df['Type 2'].astype('category')
# Convert Legendary to int data type
pokemon_df['Legendary'] = pokemon_df['Legendary'].astype('int')
pokemon_df.dtypes
```

Result:

```
Name          object
Type 1        category
Type 2        category
HP            int64
Attack        int64
Defense       int64
Sp. Atk       int64
Sp. Def       int64
Speed         int64
Generation    int64
Legendary     int64
dtype: object
```

- Using one-hot encoding to represent the attributes (Type1 and Type2)

  Use **get_dummies** function for converting the Type 1.

  ```python
  df_type1_one_hot = pd.get_dummies(pokemon_df['Type 1'])
  df_type1_one_hot.head()
  ```

  Result:

#	Bug	Dark	Dragon	Electric	Fairy	Fighting	Fire	Flying	Ghost	Grass	Ground	Ice	Normal	Poison	Psychic	Rock	Steel	Water
1	0	0	0	0	0	0	0	0	0	1	0	0	0	0	0	0	0	0
2	0	0	0	0	0	0	0	0	0	1	0	0	0	0	0	0	0	0
3	0	0	0	0	0	0	0	0	0	1	0	0	0	0	0	0	0	0
4	0	0	0	0	0	0	0	0	0	1	0	0	0	0	0	0	0	0
5	0	0	0	0	0	0	1	0	0	0	0	0	0	0	0	0	0	0

  Use **get_dummies** function for converting the Type 2.

  ```python
  df_type2_one_hot = pd.get_dummies(pokemon_df['Type 2'])
  df_type2_one_hot.head()
  ```

  Result:

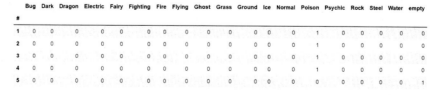

#	Bug	Dark	Dragon	Electric	Fairy	Fighting	Fire	Flying	Ghost	Grass	Ground	Ice	Normal	Poison	Psychic	Rock	Steel	Water	empty
1	0	0	0	0	0	0	0	0	0	0	0	0	0	1	0	0	0	0	0
2	0	0	0	0	0	0	0	0	0	0	0	0	0	1	0	0	0	0	0
3	0	0	0	0	0	0	0	0	0	0	0	0	0	1	0	0	0	0	0
4	0	0	0	0	0	0	0	0	0	0	0	0	0	1	0	0	0	0	0
5	0	0	0	0	0	0	0	0	0	0	0	0	0	0	0	0	0	0	1

- Combine two sets of one-hot encoding

```
# Combine Type1 and Type2 - One-hot Encoding
combine_df_one_hot = df_type1_one_hot.add(df_type2_one_hot,
                                          fill_value=0).astype('int64')
# Set the number of display columns to 30
pd.options.display.max_columns = 30


pokemon_df = pokemon_df.join(combine_df_one_hot)
pokemon_df.head()
```

Result:

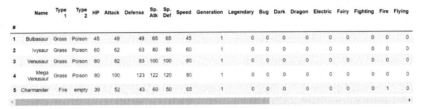

#	Name	Type 1	Type 2	HP	Attack	Defense	Sp. Atk	Sp. Def	Speed	Generation	Legendary	Bug	Dark	Dragon	Electric	Fairy	Fighting	Fire	Flying
1	Bulbasaur	Grass	Poison	45	49	49	65	65	45	1		0	0	0	0	0	0	0	0
2	Ivysaur	Grass	Poison	60	62	63	80	80	60	1		0	0	0	0	0	0	0	0
3	Venusaur	Grass	Poison	80	82	83	100	100	80	1		0	0	0	0	0	0	0	0
4	Mega Venusaur	Grass	Poison	80	100	123	122	120	80	1		0	0	0	0	0	0	0	0
5	Charmander	Fire	empty	39	52	43	60	50	65	1		0	0	0	0	0	0	1	0

- Using integer-encoding technique to represent the attributes (Type1 and Type2): convert attributes of Pokémon into numerical values (0, 1, 2, …18)

```
dict(enumerate(pokemon_df['Type 2'].cat.categories))
```

Result:

```
{0:      'Bug',
 1:      'Dark',
 2:      'Dragon',
 3:      'Electric',
 4:      'Fairy',
 5:      'Fighting',
 6:      'Fire',
 7:      'Flying',
 8:      'Ghost',
 9:      'Grass',
 10:     'Ground',
 11:     'Ice',
 12:     'Normal',
 13:     'Poison',
 14:     'Psychic',
 15:     'Rock',
 16:     'Steel',
 17:     'Water',
 18:     'empty'}
```

- The encoding value of the attributes can be received through "cat.codes"

```
pokemon_df['Type 2'].cat.codes.head(10)
```

Result:

```
1      13
2      13
3      13
4      13
5      18
6      18
7      7
8      2
9      7
10     18
dtype: int8
```

- Use numerical representation (0, 1, 2...18) to replace the original label of attributes

```
pokemon_df['Type 1'] = pokemon_df['Type 1'].cat.codes
pokemon_df['Type 2'] = pokemon_df['Type 2'].cat.codes
pokemon_df.head()
```

Result:

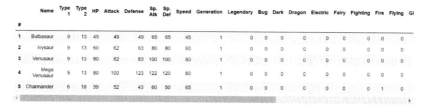

#	Name	Type 1	Type 2	HP	Attack	Defense	Sp. Atk	Sp. Def	Speed	Generation	Legendary	Bug	Dark	Dragon	Electric	Fairy	Fighting	Fire	Flying	Gl
1	Bulbasaur	9	13	45	49	49	65	65	45	1	0	0	0	0	0	0	0	0	0	0
2	Ivysaur	9	13	60	62	63	80	80	60	1	0	0	0	0	0	0	0	0	0	0
3	Venusaur	9	13	80	82	83	100	100	80	1	0	0	0	0	0	0	0	0	0	0
4	Mega Venusaur	9	13	80	100	123	122	120	80	1	0	0	0	0	0	0	0	0	0	0
5	Charmander	6	18	39	52	43	60	50	65	1	0	0	0	0	0	0	0	1	0	0

- Remove unused data (name)

```
pokemon_df.drop('Name', axis='columns', inplace=True)
pokemon_df.head()
```

Result:

#	Type 1	Type 2	HP	Attack	Defense	Sp. Atk	Sp. Def	Speed	Generation	Legendary	Bug	Dark	Dragon	Electric	Fairy	Fighting	Fire	Flying	Ghost	Grass
1	9	13	45	49	49	65	65	45	1	0	0	0	0	0	0	0	0	0	0	1
2	9	13	60	62	63	80	80	60	1	0	0	0	0	0	0	0	0	0	0	1
3	9	13	80	82	83	100	100	80	1	0	0	0	0	0	0	0	0	0	0	1
4	9	13	80	100	123	122	120	80	1	0	0	0	0	0	0	0	0	0	0	1
5	6	18	39	52	43	60	50	65	1	0	0	0	0	0	0	0	1	0	0	0

- Using 0 and 1 to represent the winner in the Pokémon combat data: 0 means the winner is the first Pokémon, 1 means the winner is the second Pokémon.

```
combats_df['Winner'] = combats_df.apply(lambda x: 0
                        if x.Winner == x.First_pokemon else 1,
                                axis='columns')
combats_df.head()
```

Result:

	First_pokemon	Second_pokemon	Winner
0	266	298	1
1	702	701	1
2	191	668	1
3	237	683	1
4	151	231	0

(c) Data division

For training and testing network model, the Pokémon dataset is divided into three sets, including training data, validation data, and test data with ratios of 6, 2, and 2, respectively. The source code for data division is as follows.

```
data_num = combats_df.shape[0]
# Get a random index equal to the number of data

indexes = np.random.permutation(data_num)
# Randomly divide data into Train, validation and test, here the division ratio is 6:2:2
train_indexes = indexes[:int(data_num *0.6)]

val_indexes = indexes[int(data_num *0.6):int(data_num *0.8)]

test_indexes = indexes[int(data_num *0.8):]

train_data = combats_df.loc[train_indexes]

val_data = combats_df.loc[val_indexes]

test_data = combats_df.loc[test_indexes]
```

- Normalize Type1 and Type 2 to limit each element between 0 and 1 (there are 19 attributes including the empty attribute)

```
pokemon_df['Type 1'] = pokemon_df['Type 1'] / 19
pokemon_df['Type 2'] = pokemon_df['Type 2'] / 19
```

- Use standard score to standardize the values of characteristics of Pokémon

```
mean = pokemon_df.loc[:, 'HP':'Generation'].mean()
std = pokemon_df.loc[:, 'HP':'Generation'].std()
pokemon_df.loc[:,'HP':'Generation'] = (pokemon_df.loc[:,'HP':'Generation']-mean)/std
pokemon_df.head()
```

Result:

#	Type 1	Type 2	HP	Attack	Defense	Sp. Atk	Sp. Def	Speed	Generation	Legendary	Bug	Dark	Dragon	Electric	Fairy	Fightir
1	0.473684	0.684211	-0.950032	-0.924328	-0.796655	-0.238981	-0.248033	-0.801002	-1.398762	0	0	0	0	0	0	
2	0.473684	0.684211	-0.362595	-0.523803	-0.347700	0.219422	0.290974	-0.284837	-1.398762	0	0	0	0	0	0	
3	0.473684	0.684211	0.420654	0.092390	0.293865	0.830626	1.009651	0.403383	-1.398762	0	0	0	0	0	0	
4	0.473684	0.684211	0.420654	0.646964	1.576395	1.502951	1.728328	0.403383	-1.398762	0	0	0	0	0	0	
5	0.315789	0.947368	-1.185007	-0.831899	-0.989065	-0.391542	-0.787041	-0.112782	-1.398762	0	0	0	0	0	0	

Convert training data in Numpy array format.

- Training, validation, and test samples from the combat data

```
x_train_index = np.array(train_data.drop('Winner', axis='columns'))
x_val_index = np.array(val_data.drop('Winner', axis='columns'))
x_test_index = np.array(test_data.drop('Winner', axis='columns'))
print(x_train_index)
```

Result: [[115 674]
　　　　[658 549]
　　　　[434 87]
　　　　...
　　　　[732 607]
　　　　[239 608]
　　　　[298 742]]

- Ground-truth labels from the combat data

```
y_train = np.array(train_data['Winner'])
y_val = np.array(val_data['Winner'])
y_test = np.array(test_data['Winner'])
```

- Prepare two different input data

   The first type: Pokémon attributes (Type1 and Type2) are represented by using integer-encoding technique.

```
# Get Pokemon's characteristics (10 characteristics: Type1, Type2, HP ...to Legendary)
pokemon_data_normal = np.array(pokemon_df.loc[:, : 'Legendary'])
print(pokemon_data_normal.shape)
# Generate input data
x_train_normal = pokemon_data_normal[x_train_index -1].reshape((-1, 20))
x_val_normal = pokemon_data_normal[x_val_index -1].reshape((-1, 20))
x_test_normal = pokemon_data_normal[x_test_index -1].reshape((-1, 20))
print(x_train_normal.shape)
```

   Result: (800, 10)

   (30000, 20)

   The second type: Pokémon attributes are represented by using one-hot encoding.

```
# Get Pokemon's characteristics HP, attack,...Legendary, bug, dark,... (Type1 and Type2
are represented by one hot encoding)
pokemon_data_one_hot = np.array(pokemon_df.loc[:, 'HP':])
print(pokemon_data_one_hot.shape)
# Generate input data
x_train_one_hot = pokemon_data_one_hot[x_train_index -1].reshape((-1, 54))
x_val_one_hot = pokemon_data_one_hot[x_val_index -1].reshape((-1, 54))
x_test_one_hot = pokemon_data_one_hot[x_test_index -1].reshape((-1, 54))
print(x_train_one_hot.shape)
```

   Result: (800, 27)

   (30000, 54)

2. Building and Training Network Models

   Two Pokémon combat prediction models are implemented including:
   (1) Model-1: Using the integer-encoding technique to represent 18 attributes (Type1 and Type2) of Pokémon for training.
   (2) Model-2: Using the one-hot encoding technique to represent 18 attributes (Type1 and Type2) of Pokémon for training.

   Table 3.5 shows the architecture of Model-1 and Model-2.

   (a) Model-1
   The input of Model-1 is the data of two Pokémon; each Pokémon has ten different characteristics, so the input shape is (20,).

TABLE 3.5   The architecture of Pokémon combat prediction models.

Name	Architecture	Description
Model-1	- Input layer with input shape (20,) - Four hidden fully connected layers: first three layers have 64 neurons and the last layer has 16 neurons - Randomly discard 30% of neurons in each of the hidden layers - Each hidden layer is followed by an ReLu activation function - Output fully connected layer with one neuron	The integer-encoding technique is applied for repressing 18 attributes of Pokémon
Model-2	- Input layer with input shape (54,) - Four hidden fully connected layers: first three layers have 64 neurons and the last layer has 16 neurons - Randomly discard 30% of neurons in each of the hidden layers - Each hidden layer is followed by an ReLu activation function - Output fully connected layer with one neuron	The one-hot encoding technique is applied for repressing 18 attributes of Pokémon.

```
inputs = keras.Input(shape=(20, ))
x = layers.Dense(64, activation='relu')(inputs)
x = layers.Dropout(0.3)(x)
x = layers.Dense(64, activation='relu')(x)
x = layers.Dropout(0.3)(x)
```

```
x = layers.Dense(64, activation='relu')(x)
x = layers.Dropout(0.3)(x)
x = layers.Dense(16, activation='relu')(x)
x = layers.Dropout(0.3)(x)
outputs = layers.Dense(1, activation='sigmoid')(x)


model_1 = keras.Model(inputs, outputs, name='model-1')
# Show network architecture
model_1.summary()
```

Result:

```
Model: "model-1"

Layer (type)                 Output Shape              Param #
=================================================================
input_1 (InputLayer)         [(None, 20)]              0

dense (Dense)                (None, 64)                1344

dropout (Dropout)            (None, 64)                0

dense_1 (Dense)              (None, 64)                4160

dropout_1 (Dropout)          (None, 64)                0

dense_2 (Dense)              (None, 64)                4160

dropout_2 (Dropout)          (None, 64)                0

dense_3 (Dense)              (None, 16)                1040

dropout_3 (Dropout)          (None, 16)                0

dense_4 (Dense)              (None, 1)                 17
=================================================================
Total params: 10,721
Trainable params: 10,721
Non-trainable params: 0
```

- Set the optimizer, loss function, and metric function

```
model_1.compile(keras.optimizers.Adam(),
                loss=keras.losses.BinaryCrossentropy(),
                metrics=[keras.metrics.BinaryAccuracy()])
```

- Create a storage directory for saving model

```
model_dir = 'lab3-logs/models'
os.makedirs(model_dir)
```

- Set the callback function

```
# Save training records as TensorBoard log files
```

```
log_dir = os.path.join('lab3-logs', 'model-1')
model_cbk = keras.callbacks.TensorBoard(log_dir=log_dir)
# Save the best model
model_mckp = keras.callbacks.ModelCheckpoint(model_dir + '/Best-model-1.h5',
                monitor='val_binary_accuracy',
                save_best_only=True,
                mode='max')
```

- Training model-1

```
history_1 = model_1.fit(x_train_normal, y_train,
                batch_size=64,
                epochs=200,
                validation_data=(x_val_normal, y_val),
                callbacks=[model_cbk, model_mckp])
```

Result:

```
Epoch 195/200
30000/30000 [==============================] - 2s 51us/sample - loss: 0.1413 - binary_accuracy: 0.9506 - val_loss: 0.1627
- val_binary_accuracy: 0.9466
Epoch 196/200
30000/30000 [==============================] - 2s 52us/sample - loss: 0.1451 - binary_accuracy: 0.9489 - val_loss: 0.1639
- val_binary_accuracy: 0.9472
Epoch 197/200
30000/30000 [==============================] - 2s 50us/sample - loss: 0.1422 - binary_accuracy: 0.9503 - val_loss: 0.1644
- val_binary_accuracy: 0.9455
Epoch 198/200
30000/30000 [==============================] - 2s 50us/sample - loss: 0.1432 - binary_accuracy: 0.9497 - val_loss: 0.1657
- val_binary_accuracy: 0.9462
Epoch 199/200
30000/30000 [==============================] - 2s 50us/sample - loss: 0.1428 - binary_accuracy: 0.9496 - val_loss: 0.1655
- val_binary_accuracy: 0.9469
Epoch 200/200
30000/30000 [==============================] - 2s 50us/sample - loss: 0.1422 - binary_accuracy: 0.9499 - val_loss: 0.1630
- val_binary_accuracy: 0.9465
```

Supplementary explanation

The BCE introduced earlier included a built-in sigmoid activation function, but the sigmoid activation function is often added to the output layer of the network model. Therefore, to avoid the output of doing two sigmoid operations, the "keras.losses.BinaryCrossentropy" function has the from_logits parameter for setting.

- **from_logits** is set to False (default): Sigmoid activation function will not be added to the loss function.
- **from_logits** is set to True: Sigmoid activation function will be added to the loss function.

The function "keras.losses.CategoricalCrossentropy" has the same concept.

(b) Model-2

The input of Model-2 is the data of two Pokémon; each Pokémon has twenty-seven different items of information, so the input shape is (54,).

```
inputs = keras.Input(shape=(54, ))

x = layers.Dense(64, activation='relu')(inputs)

x = layers.Dropout(0.3)(x)

x = layers.Dense(64, activation='relu')(x)

x = layers.Dropout(0.3)(x)

x = layers.Dense(64, activation='relu')(x)

x = layers.Dropout(0.3)(x)

x = layers.Dense(16, activation='relu')(x)

x = layers.Dropout(0.3)(x)

outputs = layers.Dense(1, activation='sigmoid')(x)


model_2 = keras.Model(inputs, outputs, name='model-2')

# Show network architecture

model_2.summary()
```

Result:

```
Model: "model-2"

Layer (type)                 Output Shape              Param #
=================================================================
input_2 (InputLayer)         [(None, 54)]              0

dense_5 (Dense)              (None, 64)                3520

dropout_4 (Dropout)         (None, 64)                0

dense_6 (Dense)              (None, 64)                4160

dropout_5 (Dropout)         (None, 64)                0

dense_7 (Dense)              (None, 64)                4160

dropout_6 (Dropout)         (None, 64)                0

dense_8 (Dense)              (None, 16)                1040

dropout_7 (Dropout)         (None, 16)                0

dense_9 (Dense)              (None, 1)                 17
=================================================================
Total params: 12,897
Trainable params: 12,897
Non-trainable params: 0
```

- Set the optimizer, loss function, and metric function

```
model_2.compile(keras.optimizers.Adam(),
        loss=keras.losses.BinaryCrossentropy(),
        metrics=[keras.metrics.BinaryAccuracy()])
```

- Set the callback function

```
# Save training records as TensorBoard log files
log_dir = os.path.join('lab3-logs', 'model-2')
model_cbk = keras.callbacks.TensorBoard(log_dir=log_dir)
# save the best model
model_mckp = keras.callbacks.ModelCheckpoint(model_dir + '/Best-model-2.h5',
                    monitor='val_binary_accuracy',
                    save_best_only=True,
                    mode='max')
```

- Training model-2

```
history_2 = model_2.fit(x_train_one_hot, y_train,
            batch_size=64,
            epochs=200,
            validation_data=(x_val_one_hot, y_val),
            callbacks=[model_cbk, model_mckp])
```

Result:

```
Epoch 195/200
30000/30000 [==============================] - 2s 51us/sample - loss: 0.0688 - binary_accuracy: 0.9715 - val_loss: 0.0980
 - val_binary_accuracy: 0.9652
Epoch 196/200
30000/30000 [==============================] - 2s 51us/sample - loss: 0.0654 - binary_accuracy: 0.9741 - val_loss: 0.1056
 - val_binary_accuracy: 0.9623
Epoch 197/200
30000/30000 [==============================] - 2s 51us/sample - loss: 0.0687 - binary_accuracy: 0.9725 - val_loss: 0.0997
 - val_binary_accuracy: 0.9637
Epoch 198/200
30000/30000 [==============================] - 2s 51us/sample - loss: 0.0686 - binary_accuracy: 0.9737 - val_loss: 0.1024
 - val_binary_accuracy: 0.9638
Epoch 199/200
30000/30000 [==============================] - 2s 51us/sample - loss: 0.0687 - binary_accuracy: 0.9722 - val_loss: 0.1010
 - val_binary_accuracy: 0.9632
Epoch 200/200
30000/30000 [==============================] - 2s 51us/sample - loss: 0.0690 - binary_accuracy: 0.9725 - val_loss: 0.1027
 - val_binary_accuracy: 0.9629
```

3. Displaying results

(a) Display the training accuracy of the Model-1 and Model-2

```
plt.plot(history_1.history['binary_accuracy'], label='model-1-training')
plt.plot(history_1.history['val_binary_accuracy'], label='model-1-validation')
plt.plot(history_2.history['binary_accuracy'], label='model-2-training')
plt.plot(history_2.history['val_binary_accuracy'], label='model-2-validation')
plt.ylabel('Accuracy')
plt.xlabel('epochs')
plt.legend()
```

Result:

Fig. 3.11 shows that the accuracy on training data and validation data of Model-2 is better than that of Model-1. The results prove that using the one-hot encoding technique for binary classification is more effective than the integer-encoding technique.

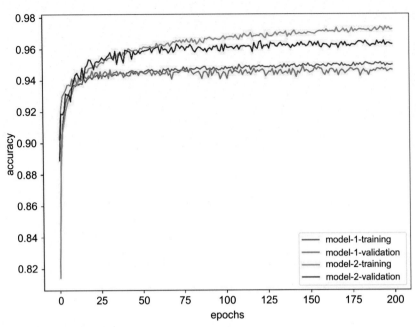

FIG. 3.11 Training result of Model-1 and Model-2.

(b) Verification on the test data

```
# Load weight of the model-1 with highest accuracy
model_1.load_weights(model_dir + '/Best-model-1.h5')
# Load weight of the model-2 with highest accuracy
model_2.load_weights(model_dir + '/Best-model-2.h5')
loss_1, accuracy_1 = model_1.evaluate(x_test_normal, y_test)
loss_2, accuracy_2 = model_2.evaluate(x_test_one_hot, y_test)
print("Model-1: {}%\nModel-2: {}%".format(accuracy_1, accuracy_2))
```

Result:

10000/10000 [==============================] - 0s 36us/sample - loss:
0.1593 - binary_accuracy: 0.9466
10000/10000 [==============================] - 0s 36us/sample - loss:
0.0947 - binary_accuracy: 0.9654
Model-1: 0.9466000199317932%
Model-2: 0.965399980545044%

As can be observed, the accuracy of Model-2 on the test data is better than that of Model-1.

(c) Pokemon PK
Finally, the trained Model-2 is used to predict the outcome of combats between three Pokémon including wonderful frog flower, spitfire dragon, and water arrow turtle, as shown in Fig. 3.12.

FIG. 3.12 Pokémon spitfire dragon, wonderful frog flower, and water arrow turtle.

- Read individual data:

```
venusaur = np.expand_dims(pokemon_data_one_hot[3], axis=0)  # Pokemon Wonderful
frog flower
charizard = np.expand_dims(pokemon_data_one_hot[7], axis=0)
#Pokemon Spitfire Dragon
blastoise = np.expand_dims(pokemon_data_one_hot[12], axis=0)
#Pokemon Water Arrow Turtle
```

- The prediction

```
# Wonderful frog flower vs Spitfire Dragon
pred = model_2.predict(np.concatenate([venusaur, charizard], axis=-1))
winner = ' Wonderful frog flower ' if pred < 0.5 else ' Spitfire Dragon '
print("pred={}, {} wins".format(pred, winner))


# Spitfire Dragon vs Water Arrow Turtle
pred = model_2.predict(np.concatenate([charizard, blastoise], axis=-1))
winner = ' Spitfire Dragon' if pred < 0.5 else ' Water Arrow Turtle '
print("pred={}, {} wins".format(pred, winner))


# Water arrow turtle vs wonderful frog flower
pred = model_2.predict(np.concatenate([blastoise, venusaur], axis=-1))
winner = ' Water Arrow Turtle ' if pred < 0.5 else ' Wonderful frog flower'
    print("pred={}, {} wins".format(pred, winner))
```

Result:

    pred=[[1.]], Spitfire Dragon wins

    pred=[[1.0699459e-07]], Spitfire Dragon wins

    pred=[[0.9999981]], Wonderful frog flower wins

The result shows that the Pokémon spitfire dragon is the strongest, followed by Pokémon wonderful frog flower and Pokémon water arrow turtle, respectively.

References

[1] J.M. Keller, M.R. Gray, J.A. Givens, A fuzzy k-nearest neighbor algorithm, in: IEEE Transactions on Systems, 1985, pp. 580–585.

[2] T. Denoeux, A k-nearest neighbor classification rule based on Dempster-Shafer theory, IEEE Trans. Syst. Man Cybern. 25 (5) (1995) 804–813.

[3] K. Fukunaga, P.M. Narendra, A branch and bound algorithm for computing k-nearest neighbors, IEEE Trans. Comput. C-24 (7) (1975) 750–753.

[4] S.A. Dudani, The distance-weighted k-nearest-neighbor rule, IEEE Trans. Syst. Man Cybern. SMC-6 (4) (1976) 325–327.

[5] J.R. Quinlan, Induction of decision trees, Mach. Learn. 1 (1986) 81–106.

[6] H. Schmid, Probabilistic part of speech tagging using decision trees, in: Proceedings of the International Conference on New Methods in Language Processing, 1994, pp. 44–49.

[7] C.Z. Janikow, Fuzzy decision trees: issues and methods, IEEE Trans. Syst. Man Cybern. B 28 (1) (1998) 1–14.

[8] S. Tsang, B. Kao, K.Y. Yip, W. Ho, S.D. Lee, Decision trees for uncertain data, IEEE Trans. Knowl. Data Eng. 23 (1) (2011) 64–78.

[9] L. Breiman, Random forests, Mach. Learn. 45 (1) (2001) 5–32.

[10] A. Bosch, A. Zisserman, X. Munoz, Image classification using random forests and ferns, in: IEEE International Conference on Computer Vision, Rio de Janeiro, 2007, pp. 1–8.

[11] P.O. Gislason, J.A. Benediktsson, J.R. Sveinsson, Random forests for land cover classification, Pattern Recogn. Lett. 27 (4) (2006) 294–300.

[12] C. Cortes, V. Vapnik, Support-vector network, Mach. Learn. (1995) 273–297.

[13] T. Joachims, Text categorization with support vector machines: learning with many relevant features, in: European Conference on Machine Learning, 1998, pp. 137–142.

[14] W.M. Campbell, D.E. Sturim, D.A. Reynolds, Support vector machines using GMM supervectors for speaker verification, IEEE Signal Process. Lett. 13 (5) (2006) 308–311.

[15] K.P. Bennett, A. Demiriz, Semi-supervised support vector machines, in: Advances in Neural Information Processing Systems, 1999, pp. 368–374.

[16] T. Le, P. Lin, S. Huang, LD-Net: an efficient lightweight denoising model based on convolutional neural network, IEEE Open J. Comput. Soc. 1 (2020) 173–181.

[17] T. Le, S. Huang, D. Jaw, Cross-resolution feature fusion for fast hand detection in intelligent homecare systems, IEEE Sensors J. 19 (12) (2019) 4696–4704.

[18] Y. Liu, D. Jaw, S. Huang, J. Hwang, DesnowNet: context-aware deep network for snow removal, IEEE Trans. Image Process. 27 (6) (2018) 3064–3073.

[19] T. Le, D. Jaw, I. Lin, S. Huang, An efficient hand detection method based on convolutional neural network. International Symposium on Next Generation Electronics (ISNE), (2018) pp. 1–2, https://doi.org/10.1109/ISNE.2018.8394651.

[20] J.A. Hartigan, M.A. Wong, A. K-Means Clustering, Algorithm, Appl. Stat. 28 (1) (1979) 100–108.

[21] A.K. Jain, Data clustering: 50 years beyond K-means, Pattern Recogn. Lett. 31 (8) (2010) 651–666.

[22] T. Kanungo, et al., An efficient k-means clustering algorithm: analysis and implementation, IEEE Trans. Pattern Anal. Mach. Intell. 24 (7) (2002) 881–892.

[23] S.Z. Selim, M.A. Ismail, K-Means-type algorithms: a generalized convergence theorem and characterization of local optimality, IEEE Trans. Pattern Anal. Mach. Intell. PAMI-6 (1) (1984) 81–87.

[24] J.F. Navarro, C.S. Frenk, S.D.M. White, A universal density profile from hierarchical clustering, Astrophys. J. 490 (2) (1997) 493–508.

[25] P. Bajcsy, N. Ahuja, Location- and density-based hierarchical clustering using similarity analysis, IEEE Trans. Pattern Anal. Mach. Intell. 20 (9) (1998) 1011–1015.

[26] X. Tang, P. Zhu, Hierarchical clustering problems and analysis of fuzzy proximity relation on granular space, IEEE Trans. Fuzzy Syst. 21 (5) (2013) 814–824.

[27] S. Wold, K. Esbensen, P. Geladi, Principal component analysis, Chemom. Intell. Lab. Syst. 2 (1987) 37–52.

[28] I. Joliffe, Principal Component Analysis, second ed., Springer, New York, 2002.

[29] B. Moore, Principal component analysis in linear systems: controllability, observability, and model reduction, IEEE Trans. Autom. Control 26 (1) (1981) 17–32.

[30] M.E. Tipping, Sparse kernel principal component analysis, in: T.K. Leen, T.G. Dietterich, V. Tresp (Eds.), Advances in Neural Information Processing Systems, MIT Press, Cambridge, 2000, pp. 633–639.

[31] S. Mika, B. Scholkopf, A. Smola, K. Muller, M. Scholz, G. Ratsch, Kernel PCA and de-noising in feature spaces, Adv. Neural Inf. Proces. Syst. 11 (1) (1999) 536–542.

[32] C. Liu, Gabor-based kernel PCA with fractional power polynomial models for face recognition, IEEE Trans. Pattern Anal. Mach. Intell. 26 (5) (2004) 572–581.

[33] J.E. van Engelen, H.H. Hoos, A survey on semi-supervised learning. Mach. Learn. 109 (2) (2020) 373–440, Online Available: https://doi.org/10.1007/s10994-019-05855-6.

[34] D. Silver, et al., Mastering the game of go with deep neural networks and tree search, Nature 529 (7587) (2016) 484–489.

[35] D. Silver, et al., Mastering the game of go without human knowledge, Nature 550 (7676) (2017) 354–359.

[36] H. Qin, R. Gong, X. Liu, X. Bai, J. Song, N. Sebe, Binary neural networks: a survey. Pattern Recogn. 105 (2020) 107281, https://doi.org/10.1016/j.patcog.2020.107281.

4

Multi-category classification problem

> **OUTLINE**
> - Introduction to convolutional neural networks (CNNs)
> - Using categorical cross-entropy
> - Using data augmentation techniques to increase the training data
> - Completing multi-category classification on the CIFAR-10 dataset using a CNN

4.1 Convolutional neural network

4.1.1 Introduction to convolutional neural network

A convolutional neural network, also known as CNN or ConvNet, is a class of deep neural network that has been successfully applied to various computer vision applications, especially for analyzing visual images [1]. Fig. 4.1 shows the rank of the best methods from the 2012 ImageNet Large Scale Visual Recognition Challenge (ImageNet LSVRC [2]). As shown, AlexNet using CNN is superior to the previous state-of-the-art image classification methods in terms of error rate. In subsequent ImageNet competitions, the models based on CNN, such as GoogLeNet [3], ResNet [4], ResNeXt [5], and others, continue to be the winners and overwhelm traditional models with impressive performance.

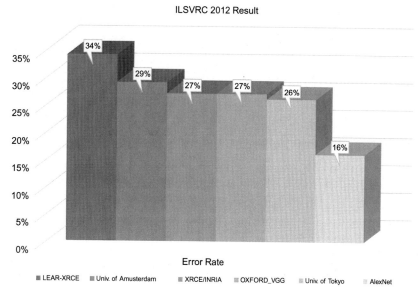

FIG. 4.1 The rank of the best methods from the 2012 ImageNet Large Scale Visual Recognition Challenge (ImageNet LSVRC).

As discussed in previous chapters, the regular neural networks include an input layer, hidden layers, and an output layer, where an input from the input layer is transformed through a series of hidden layers before being sent to the output layer. Each layer contains a set of neurons, in which each neuron is fully connected to all neurons in the previous layer, and neurons operate completely independently, as shown in Fig. 4.2A. The CNNs are similar to the regular neural networks, which means they are also made of layers containing neurons with learnable weights and biases. However, different from regular neural networks, neurons in the layers of CNNs are organized in three dimensions, including width, height, and depth, and each neuron is only connected to a local region in the previous layer through a kernel (receptive field of the neuron), as shown in Fig. 4.2B. In CNNs, the 3D input volume of neurons is transformed to a 3D output volume of neurons at every layer. For example, if a CNN designed for image classification on the ImageNet [2] dataset receives an image with dimensions $256 \times 256 \times 3$ as the input, the final output layer of the CNN should have dimensions of $1 \times 1 \times 1000$ because ImageNet has 1000 classes.

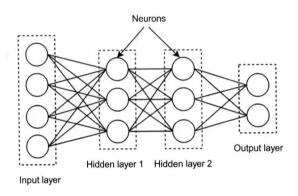

(A) A regular 3-layer neural network.

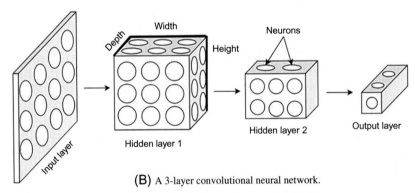

(B) A 3-layer convolutional neural network.

FIG. 4.2   Illustration of a regular neural network and a convolutional neural network (CNN).

4.1.2 Building a convolutional neural network

In general, a simple CNN is built with three main types of layers: convolutional layer, pooling layer, and fully connected layer, as shown in Fig. 4.3. The following introduces these layers in more detail.

1. Convolutional layer
   The principle of the convolution layer is to use kernels or filters to slide on the input for learning features. As shown in Fig. 4.4, a 3×3 kernel is slid onto a 4×4 input image to produce a 2×2 output feature map in size; the output of the first calculation is 7, the calculation formula is:

   $$(1 \times 1) + (1 \times 1) + (0 \times 0) +$$

$$(2 \times 0) + (3 \times (-1)) + (2 \times 2) +$$
$$(1 \times 2) + (4 \times 0) + (2 \times 1) = 7$$

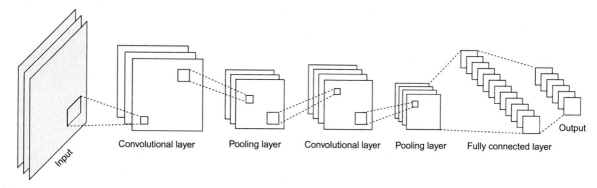

FIG. 4.3 Illustration of convolutional neural network (CNN) architecture.

The size and number of kernels are adjustable hyperparameters, and most of the sizes are set to odd numbers such as 3×3, 5×5, or 7×7. Since too many kernels may also cause over-fitting problem, it is necessary to adjust this parameter to select the most suitable ones.

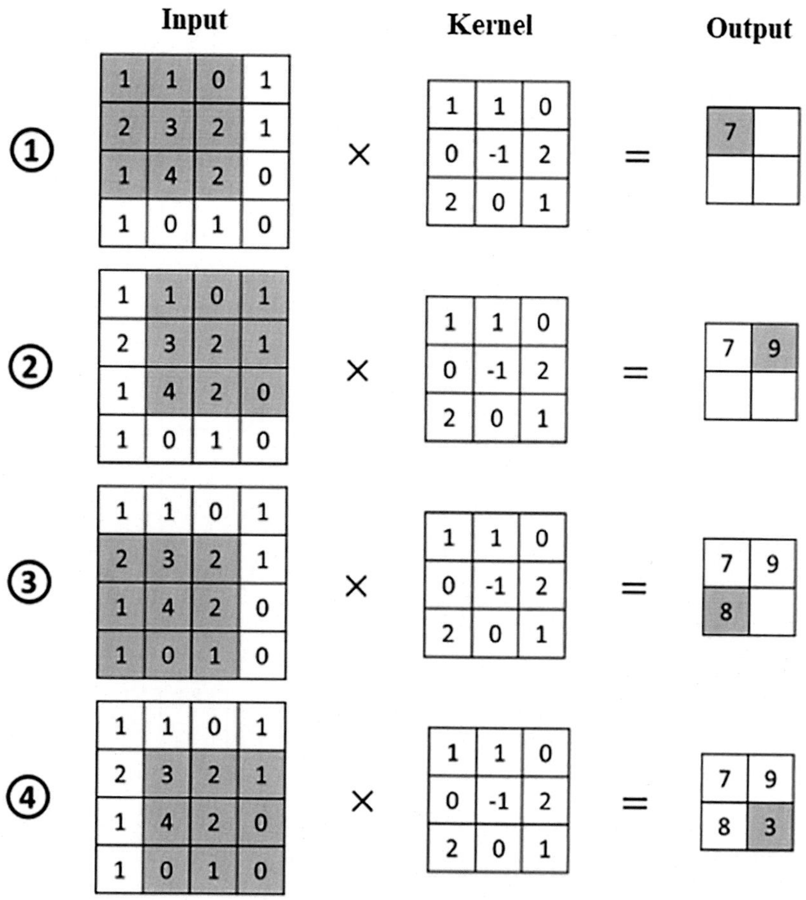

FIG. 4.4 Computation in the convolutional layer.

Zero padding: Fig. 4.4 describes the basic operation of the convolutional layer, in which after each convolution, the size of the output volume is changed compared with that of the input volume. To control the size of the output volume, a zero-padding technique is applied by padding zeros around the border of the input volume. In most cases, zero padding is used to retain the spatial size of the input volume. For example, by applying zero padding with a size of 1 around the border of a 4×4 input image, the input size becomes 6×6, as shown in Fig. 4.5. Then, performing convolution with a kernel size of 3×3 on the input with a size of 6×6, the spatial size of the output volume is 4×4, which is the same size as the original input, as shown in Fig. 4.6.

Input   **Output**

1	1	0	1
2	3	2	1
1	4	2	0
1	0	1	0

Zero-padding →

0	0	0	0	0	0
0	1	1	0	1	0
0	2	3	2	1	0
0	1	4	2	0	0
0	1	0	1	0	0
0	0	0	0	0	0

FIG. 4.5   Zero padding with size of 1.

Input with Padding

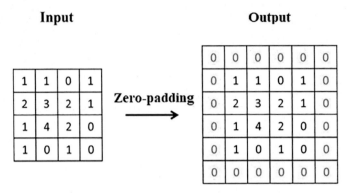

Kernel

1	1	0
0	-1	2
2	0	1

Output

4	5	9	3
9	7	9	4
9	8	3	5
0	7	5	2

FIG. 4.6   Convolution operation with zero padding.

Stride: Stride controls how the kernel slides on the input. In the general convolution layer, stride with a value of 1 is used, which means the kernel is slid one pixel at each time. When stride is greater than 1, such as stride of 2 in Fig. 4.7, the kernel is moved 2 pixels at a time, resulting in smaller output volume spatially.

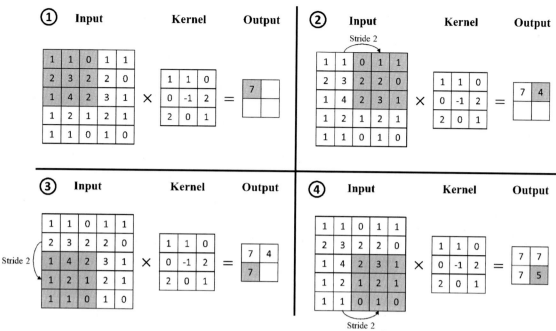

FIG. 4.7 Convolution operation with stride 2.

The formulas for calculating the spatial size of the output of the convolution layer in two padding modes of same padding mode and valid padding mode are expressed as follows.

- Same padding (padding): The spatial size of the output is the same size as that of the input after convolution

$$Output_{height} = \frac{Input_{height}}{Stride.}$$

$$Output_{width} = \frac{Input_{width}}{Stride}$$

- Valid padding (no padding): The spatial size of the output is smaller than that of the input after convolution

$$Output_{height} = \frac{Input_{height} - kernel_{height} + 1}{Stride}$$

$$Output_{width} = \frac{Input_{width} - kernel_{width} + 1}{Stride}$$

Input: Input size
Output: Output size
kernel: kernel size or convolution filter size
Stride: stride size of the convolution operation

The preceding examples are the computations of a single input channel and single kernel, but in fact, in the convolutional layer, the computation is performed on multiple input channels and multiple kernels. For example, the convolution is performed on the color input image, as shown in Fig. 4.8, in which the parameters are set as:

- Input image: $4 \times 4 \times 3$ (height, width, depth), here the depth ($Input_{channel}$) is shown as three colors: R (Red), G (Green), and B (Blue).
- Padding: valid
- Stride: 1
- Kernel number ($kernel_{numbers}$): 2 (W0 and W1)
- Kernel size: 3×3 ($kernel_{height} \times kernel_{width}$)
- Bias: None

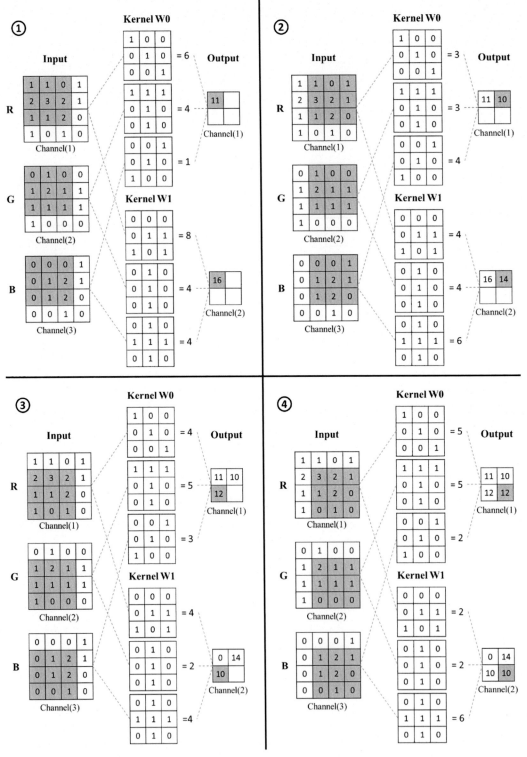

FIG. 4.8 Convolution operation on an image with size of $4 \times 4 \times 3$ with two kernels W0 ($3 \times 3 \times 3$) and W1 ($3 \times 3 \times 3$).

The number of parameters of the convolutional layer: The formula for calculating the number of parameter of the convolution layer is:

$$Parameter = \left(Input_{channel} \times kernel_{height} \times kernel_{width} + Bias \right) \times kernel_{numbers}$$

$Input_{channel}$: The depth or the number of channels of the input
$kernel_{height}$: The height of the kernel

$kernel_{width}$: The width of the kernel
$kernel_{numbers}$: The number of kernels
Bias: if bias is used, bias $= 1$, otherwise bias $= 0$

Example 1: Calculating the number of parameters of the convolutional layer in Fig. 4.8.

$$Parameter = (3 \times 3 \times 3 + 0) \times 2 = 54$$

Example 2: Create a CNN that comprised of an input layer with size of $28 \times 28 \times 4$ and an output convolutional layer using 32 kernels, each kernel with size of 3×3. The parameter settings of the network are:

- Input: $28 \times 28 \times 4$ (height, width, and depth)
- Padding: valid
- Stride: 1
- Kernel number: 32 ($kernel_{numbers}$)
- Kernel size: 3×3 ($kernel_{height}$, $kernel_{width}$)
- Bias: Yes

Source code for building the CNN:

```
from tensorflow import keras

inputs = keras.Input((28, 28, 4))

outputs = keras.layers.Conv2D(32, kernel_size=3, strides=(1, 1), padding='valid',
        use_bias=True)(inputs)

model = keras.Model(inputs, outputs)

model.summary()
```

The number of parameters of the CNN can be obtained through the "model.summary()" function; it is 1184 parameters, as shown in Fig. 4.9.

```
Model: "model"

_____
 Layer (type)                Output Shape              Param #
=================================================================
 input_1 (InputLayer)        [(None, 28, 28, 4)]       0

 conv2d (Conv2D)             (None, 26, 26, 32)         1184
=================================================================
Total params: 1,184
Trainable params: 1,184
Non-trainable params: 0
```

FIG. 4.9 Summary of the convolutional neural network (CNN).

To verify whether the number of parameters of the CNN is the same as the 1184 shown in Fig. 4.9, the parameter calculation formula above is applied. Because the CNN has only one convolutional layer, the number of parameters of the network is the number of parameters of the convolutional layer, computed as:

$$Parameter = (4 \times 3 \times 3 + 1) \times 32 = 1184$$

As shown, the calculation result is consistent with the number of parameters computed through the "model.summary ()" function in Fig. 4.9.

Supplementary explanation

Because the vanishing gradients problem is easy to encounter during training when using the sigmoid activation function in convolutional layers of the CNN, this activation function is replaced with the rectified linear activation (ReLU) function in current CNNs for effective training. Chapter 5 provides more information on the impact of sigmoid function and ReLU function on the performance of CNNs.

2. Pooling layer

The pooling layer is often adopted immediately after the convolution layer to reduce the spatial size (width and height) of the input volume for reducing the computation in the CNNs and avoiding the overfitting problem during the training process. The pooling layer works independently with every channel of the input, therefore the number of the channels or depth dimension of the presentation remains unchanged, as shown in Fig. 4.10. There are some pooling layers such as max pooling, average pooling, and L2 norm pooling layers. Among them, the max pooling layer, which adopts max operation, is most commonly used in CNNs since it has proven to work more effectively in practice. An example in Fig. 4.11 is the operation of the max pooling layer, where the input of size 4×4 is pooled with a 2×2 kernel and stride 2, resulting in an output of size 2×2.

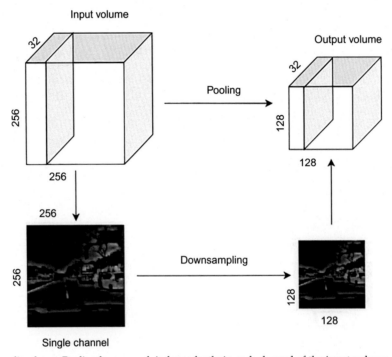

FIG. 4.10   Illustration of pooling layer. Pooling layers work independently in each channel of the input volume.

3. Fully connected layer

The fully connected layer or dense layer is an important component of CNNs. This layer has neurons with full connection to all neurons in the previous layer. It has been applied successfully in many computer vision applications such as image classification, semantic segmentation, and so on. In CNNs, convolution and pooling operations are performed first through convolutional layers and pooling layers to extract features from the input. The resulting features are flattened into a one-dimensional feature vector before being sent to the fully connected layer for combining data and driving the final output, as shown in Fig. 4.12. If the neural networks are only constructed by fully connected layers, they are called fully connected networks. The fully connected networks are often extremely computationally expensive and have lower performance than CNNs. In the next section, we apply both a fully connected network and a CNN for multi-category classification to explore the effectiveness of each.

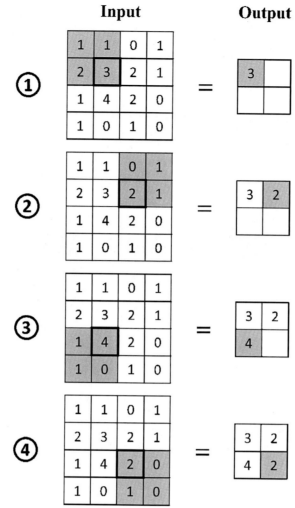

FIG. 4.11   Operation of the max pooling layer.

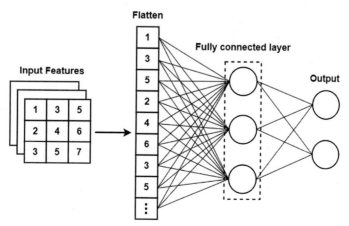

FIG. 4.12   The connection of a fully connected layer.

4.1.3 Operation of convolutional neural network

In CNNs, each convolutional layer has kernels for feature extraction from the input, and using kernels with different element values can obtain various effects, as shown in Table 4.1. The kernels have no meaning at the beginning; their values need to be initialized and updated during training process.

TABLE 4.1   The meaning of different kernel parameters.

Input image	Kernel	Output image
	$\begin{bmatrix} 0 & 0 & 0 \\ 0 & 1 & 0 \\ 0 & 0 & 0 \end{bmatrix}$	
	$\begin{bmatrix} 1 & 0 & -1 \\ 0 & 0 & 0 \\ -1 & 0 & 1 \end{bmatrix}$	
	$\begin{bmatrix} -1 & -1 & -1 \\ -1 & 8 & -1 \\ -1 & -1 & -1 \end{bmatrix}$	

The preceding description only mentions the network with one convolutional layer. How do CNNs with multiple layers work? Fig. 4.13 illustrates the operation of a deep CNN for image classification, in which the first few convolutional layers (layers 1–3) are mainly responsible for extracting simple features of the input image such as edges, lines, and so on. Then, the resulting features are sent to the deeper layers (layers 4–6) for generating more specific features such as nose, eyes, and ears of objects. Finally, these output features are flattened into one-dimensional feature vector by a flatten layer and then fed into the fully connected layer for final prediction.

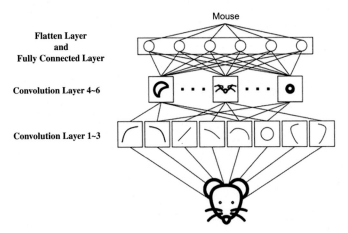

FIG. 4.13   Deep convolutional network.

Figs. 4.14–4.16 show examples of feature visualization in a CNN where the trained VGG-16 [6] network on ImageNet [2] and the DeconvNet [7] are employed for obtaining results. As shown in Fig. 4.14, the shallow layer (layer 2) of the VGG-16 network responds to conjunctions of lines and edges.

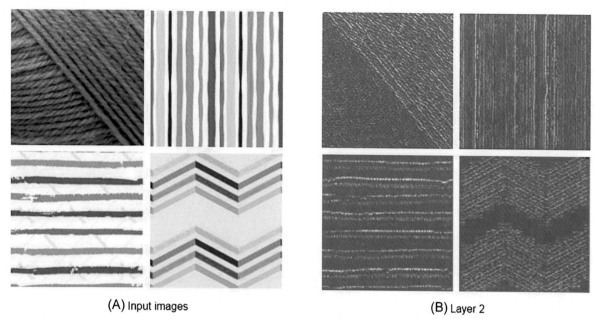

FIG. 4.14 Visualization of features in the second layer of the trained VGG-16 model.

In Fig. 4.15, the middle layer (layer 7) of the VGG-16 network contains more complex invariances, extracting similar textures such as the shape of the moon, the windows, and the house.

FIG. 4.15 Visualization of features in the seventh layer of the trained VGG-16 model.

In Fig. 4.16, the deep layer (layer 15) of the VGG-16 network presents the object with significant variation and more class specificity, such as the eyes, ears of the dogs and cat, the leg and pose of the bird, and so on.

(A) Input images   (B) Layer 15

FIG. 4.16   Visualization of features in the fifteenth layer of the trained VGG-16 model.

4.2 Multi-category classification

4.2.1 Introduction to multi-category classification

In machine learning, multi-category classification refers to the problem of categorizing samples into one of three or more classes. Similar to the binary classification models introduced in Chapter 3, the multi-category classification models are also used to model the probability of classes. However, unlike the binary classification models, which output a single probability to infer the input sample into one of two classes, the multi-category classification models produce one probability per class, in which the sum of output probability scores should be equal to 1, as shown in Fig. 4.17. Based on the predicted probability scores, the model can determine which target class that the input sample belongs to. For example, given a model classifying three classes, after feeding a sample to the model, it outputs the probability score of 0.7 for class 1, 0.1 for class 2, and 0.2 for class 3. This means the model believes with 70% confidence that the input sample is in class 1, 10% confidence it is in class 2, and 20% it is in class 3.

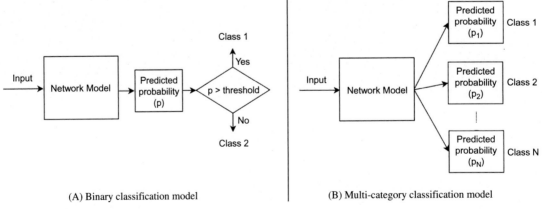

(A) Binary classification model   (B) Multi-category classification model

FIG. 4.17   Flowchart of classification models.

4.2.2 Multi-category classification model

1. Network architecture

As introduced in Section 4.1, state-of-the-art multi-category classification models are based on CNNs, which have two main components of feature extractor and classifier, as shown in Fig. 4.18.

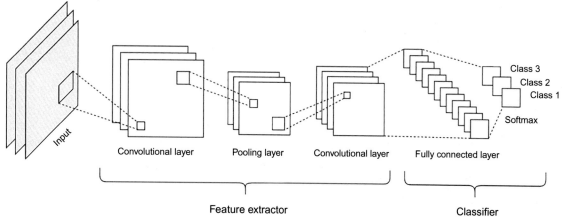

FIG. 4.18 The architecture of a CNN-based multi-category classification model.

Feature extractor: The feature extractor consists of convolutional layers and pooling layers for extracting features from the input.

Classifier: The classifier consists of fully connected layers that combine extracted features from the feature extractor for performing the final classification decision. To accomplish this goal, the last fully connected layer is followed by softmax function to compute a probability score for each target class. Described by the probability model, each output value of the softmax function is between 0 and 1, and the sum of all output values is guaranteed to be 1. The softmax function is expressed as:

$$y_i = \frac{e^{z_i}}{\sum_{j=1}^{C} e^{z_j}} \quad \text{for } i = 1, 2, ..., C \text{ and } z = (z_1, z_2, ..., z_C) \in R^C$$

For example, a model of classifying three classes ($C = 3$) outputs three values of $z_1 = 3$, $z_2 = 1$, and $z_3 = -3$ at the last layer, and these values are passed through the softmax function to convert to probability scores of $y_1 = 0.88$, $y_2 = 0.12$, and $y_3 = 0$, respectively, as shown in Fig. 4.19. Based on these output probability scores the input sample clearly belongs to the first class.

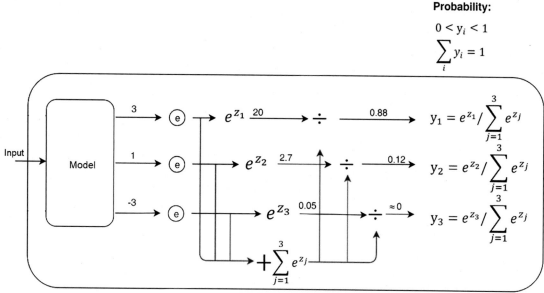

FIG. 4.19 Softmax output calculation.

2. Loss function

As mentioned in the previous chapter, the classification problem can be divided into a binary classification problem and a multi-category classification problem. While binary cross-entropy (BCE) is usually used as the loss function for binary classification models, categorical cross-entropy (CCE) is employed as the loss function for multi-category classification models. The CCE loss function is a combination of softmax function and cross-entropy (CE) function. It is also called softmax loss, as shown in Fig. 4.20.

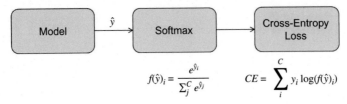

$$f(\hat{y})_i = \frac{e^{\hat{y}_i}}{\sum_j^C e^{\hat{y}_j}} \qquad CE = \sum_i^C y_i \log(f(\hat{y})_i)$$

FIG. 4.20   Categorical cross-entropy (CCE).

The formula for CE is defined:

$$CE = -\frac{\sum_{i=1}^{N}\sum_{j=1}^{C} y_{i,j} \log \hat{y}_{i,j}}{N}$$

The formula for CCE is expressed:

$$CCE = -\frac{\sum_{i=1}^{N}\sum_{j=1}^{C} y_{i,j} \log \left(f\left(\hat{y}_{i,j} \right) \right)}{N}$$

where, y is the expected output, $\hat{y}$ is the predicted value of the model, f is softmax function, C is number of categories, and N is the amount of data in a batch.

Fig. 4.21 shows an example for calculating CCE. The neural network with the output layer followed by the softmax function takes the "cat" image as the input and outputs the prediction result of [0,0,0,0.6,0,0,0.1,0,0.3,0]. The error between the expected output [0,0,0,1,0,0,0,0,0,0] and the prediction result is calculated using:

$$
\begin{aligned}
loss &= -(0 \times log\,0 + 0 \times log\,0 + 0 \times log\,0 + 1 \times log\,0.6 + 0 \times log\,0 + 0 \\
&\quad \times log\,0 + 0 \times log\,0.1 + 0 \times log\,0 + 0 \times log\,0.3 + 0 \times log\,0) \\
&= -1 \times log\,0.6 \\
&= 0.22
\end{aligned}
$$

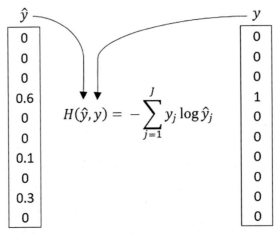

FIG. 4.21   A calculation example using categorical cross-entropy (CCE).

4.2.3 Data augmentation

As mentioned in Chapter 2, the deep learning models often encounter the problem of overfitting during training. As such, three methods are used to prevent this problem: reducing the size of the model, weight regularization, and dropout. Another way to deal with the problem of overfitting is to increase the amount and diversity of the training data by applying different transformations to the available data; this is called data augmentation [8–13]. Because data augmentation is most commonly used in image processing, it is also called image augmentation. By using image augmentation, the amount of data can be increased more than two times compared with the original data, as shown in Fig. 4.22. The common image transformations techniques are:

- image flipping
- image rotation
- image shifting
- image scaling
- color conversion (contrast, saturation or brightness, etc.)
- blur image (Gaussian blur or average blur, etc.)
- add noise (Gaussian noise or pepper and salt noise, etc.)

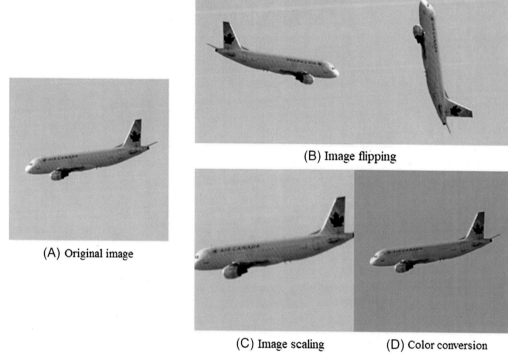

(A) Original image

(B) Image flipping

(C) Image scaling

(D) Color conversion

FIG. 4.22 Example of image augmentation.

4.3 Experiment: CIFAR-10 image classification

In this section, we use a CIFAR-10 dataset to train and evaluate the performance of the multi-category classification models. For training, we use CCE and Adam as the loss function and optimizer, respectively. The network models are implemented in three scenarios.

(1) Model-1: fully connected neural network (FCNN)
(2) Model-2: CNN without applying image augmentation technique for training the model
(3) Model-3: CNN with image augmentation technique applied to increase the amount and diversity of training data

4.3.1 Introduction to CIFAR-10 dataset

The CIFAR-10 dataset [14] contains 60,000 RGB images with size of 32×32 for each, with 50,000 images used for training and 10,000 images for testing purposes. CIFAR-10 consists of 10 classes: airplane, automobile, bird, cat, deer, dog, frog, horse, ship, and truck, as shown in Fig. 4.23. The dataset can be downloaded at the official website: www.cs.toronto.edu/~kriz/cifar.html.

FIG. 4.23 CIFAR-10 dataset.

4.3.2 TensorFlow datasets

The dataset plays a very important role in deep learning, but the different source format and complexities of datasets make it difficult to simply load them into deep learning models. To address this problem, TensorFlow provides a collection of ready-to-use public datasets called TensorFlow Datasets. In TensorFlow Datasets, all datasets are provided as "tf.data.Datasets," supporting ease of use and highly optimized input pipelines. Here we present an example of loading data for training the network model using tf.data.Datasets.

```
import tensorflow_datasets as tfds

# Load dataset

train_data = tfds.load("mnist", split= tfds.Split.TRAIN)

# Set the input pipeline

train_data = train_data.shuffle(1024).batch(32).prefetch(tf.data.experimental.AUTOTUNE)

# Train network

model.fit(train_data, epochs=100)
```

There are more than 60 datasets for image classification. Many datasets for various fields such as object detection, question answering, summarization, and so on, are available on TensorFlow Datasets. Please go to www.tensorflow.org/datasets/ for more details. In the next section, we load the CIFAR10 dataset from TensorFlow Datatsets through "tf.data.Datasets" API for training and testing multi-category classification models.

4.3.3 Code examples

Fig. 4.24 is the flowchart of the source code for multi-category classification models.

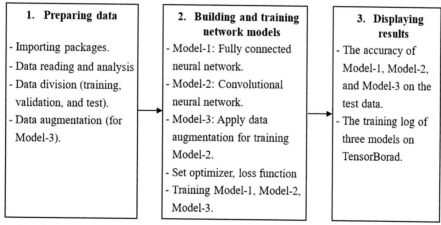

FIG. 4.24 Flowchart of the source code for multi-category classification models.

1. Preparing data
   (a) Import packages

```
        import os
import numpy as np
import pandas as pd
import tensorflow as tf
import matplotlib.pyplot as plt
from tensorflow import keras
from tensorflow.keras import layers
```

```
import tensorflow_datasets as tfds
```

(b) Data reading and analysis

■ View the current datasets provided by TensorFlow Datasets:

```
tfds.list_builders()
```

Result:

```
['abstract_reasoning',
 'bair_robot_pushing_small',
 'caltech101',
 'cats_vs_dogs',
 'celeb_a',
 'celeb_a_hq',
 'chexpert',
 'cifar10',
 'cifar100',
 'cifar10_corrupted',
 'cnn_dailymail',
 'coco2014',
 'colorectal_histology',
 'colorectal_histology_large',
 'cycle_gan',
 'diabetic_retinopathy_detection',
 'dsprites',
 'dtd',
 'dummy_dataset_shared_generator',
 'dummy_mnist',
 'emnist',
 'fashion_mnist',
 'flores',
 'glue',
 'groove',
 'higgs',
 'horses_or_humans',
 'image_label_folder',
 'imagenet2012',
 'imagenet2012_corrupted',
 'imdb_reviews',
 'iris',
 'kmnist',
 'lm1b',
 'lsun',
 'mnist',
 'moving_mnist',
 'multi_nli',
 'nsynth',
 'omniglot',
 'open_images_v4',
 'oxford_flowers102',
 'oxford_iiit_pet',
 'para_crawl',
 'quickdraw_bitmap',
 'rock_paper_scissors',
 'shapes3d',
 'smallnorb',
```

```
                        'squad',
                        'starcraft_video',
                        'sun397',
                        'svhn_cropped',
                        'ted_hrlr_translate',
                        'ted_multi_translate',
                        'tf_flowers',
                        'titanic',
                        'ucf101',
                        'voc2007',
                        'wikipedia',
                        'wmt15_translate',
                        'wmt16_translate',
                        'wmt17_translate',
                        'wmt18_translate',
                        'wmt19_translate',
                        'wmt_translate',
                        'xnli']
```

- Load CIFAR-10 dataset:

```
# Divide data into training data and validation data with a ratio of 9:1

train_split, valid_split = ['train[:90%]', 'train[90%:]']
# Get training data

train_data, info = tfds.load("cifar10", split=train_split, with_info=True)
# Get valid data

valid_data = tfds.load("cifar10", split=valid_split)
# Get test data

test_data = tfds.load("cifar10", split=tfds.Split.TEST)
```

- Display CIFAR-10 information

Some basic information of the dataset such as input image size, number of classes, the number of samples of training set, and so on, can display through the following simple command:

```
print(info)
```

Result:

```
tfds.core.DatasetInfo(
    name='cifar10',
    version=1.0.2,
    description='The CIFAR-10 dataset consists of 60000 32x32 colour images in 10 classes, with 6000
images per class. There are 50000 training images and 10000 test images.',
    urls=['https://www.cs.toronto.edu/~kriz/cifar.html'],
    features=FeaturesDict({
        'image': Image(shape=(32, 32, 3), dtype=tf.uint8),
        'label': ClassLabel(shape=(), dtype=tf.int64, num_classes=10)
```

```
        },
        total_num_examples=60000,
        splits={
            'test': <tfds.core.SplitInfo num_examples=10000>,
            'train': <tfds.core.SplitInfo num_examples=50000>
        },
        supervised_keys=('image', 'label'),
        citation="""
            @TECHREPORT{Krizhevsky09learningmultiple,
                author = {Alex Krizhevsky},
                title = {Learning multiple layers of features from tiny images},
                institution = {},
                year = {2009}
            }

        """,
        redistribution_info=,
    )
```

- Display 10 classes of CIFAR-10

```
labels_dict = dict(enumerate(info.features['label'].names))

labels_dict
```

Result:

```
{0: 'airplane',
 1: 'automobile',
 2: 'bird',
 3: 'cat',
 4: 'deer',
 5: 'dog',
 6: 'frog',
 7: 'horse',
 8: 'ship',
 9: 'truck'}
```

- View training data and calculate the number of each class

```
# Create a dict to count the number of tags in each category

train_dict = {}

# Read the entire training data set

for data in train_data:

    # Convert the read label to numpy format

    label = data['label'].numpy()

    # Count the number of each category: use a dictionary

    train_dict[label] = train_dict.setdefault(label, 0) + 1

print(train_dict)
```

Result: {0: 4492, 1: 4473, 2: 4491, 3: 4497, 4: 4481, 5: 4519, 6: 4509, 7: 4515, 8: 4517, 9: 4506}

- Display images

```
# Create an array to display images
output = np.zeros((32 * 4, 32 * 4, 3), dtype=np.uint8)
row = 0
for data in train_data.batch(4).take(4):
    output[:, row*32:(row+1)*32] = np.vstack(data['image'].numpy())
    row += 1
# Set the display window size
plt.figure(figsize=(4, 4))
# Display image
plt.imshow(output)
```

Result:

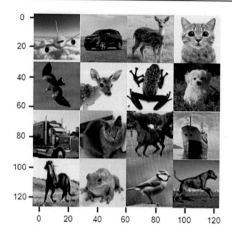

(c) Data division
  - Converting data
    - Normalization: divide all pixels in the image by 255 to scale the pixel value between 0 and 1
    - Label data: using one-hot encoding, for example, category 2 is represented as [0,0,0,0,0,0,0,0,1,0]

```
def parse_fn(dataset):
    # Image standardization
    x = tf.cast(dataset['image'], tf.float32) / 255.
    # Convert the output label to One-hot encoding
    y = tf.one_hot(dataset['label'], 10)
    return x, y
```

- Setting training data, validation data, and test data

```
AUTOTUNE = tf.data.experimental.AUTOTUNE  # Automatic adjustment mode
batch_size = 64  # Batch size
train_num = int(info.splits['train'].num_examples / 10) * 9  # Number of training


# Shuffle training data
train_data = train_data.shuffle(train_num)
# Training data
train_data = train_data.map(map_func=parse_fn, num_parallel_calls=AUTOTUNE)
# Set the batch size to 64 and turn on prefetch mode
train_data = train_data.batch(batch_size).prefetch(buffer_size=AUTOTUNE)


# Validation data
valid_data = valid_data.map(map_func=parse_fn, num_parallel_calls=AUTOTUNE)
# Set the batch size to 64 and turn on prefetch mode
valid_data = valid_data.batch(batch_size).prefetch(buffer_size=AUTOTUNE)


# Test data
test_data = test_data.map(map_func=parse_fn, num_parallel_calls=AUTOTUNE)
# Set the batch size to 64 and turn on prefetch mode
test_data = test_data.batch(batch_size).prefetch(buffer_size=AUTOTUNE)
```

(d) Data Augmentation (for model-3)
- Building help functions
  - Read image

```
x = 3
y = 7
# Read image
image_test = output[y*32:(y+1)*32, x*32:(x+1)*32, :]
# Display image
plt.imshow(image_test)
```

Result:

- Flip horizontally

```
def flip(x):
    """
    flip image
    """
    x = tf.image.random_flip_left_right(x)    # # Randomly flip the image
    return x
image_2 = flip(image_test)
image = np.hstack((image_test, image_2))
# Display image
plt.imshow(image)
```

Result:

- Color conversion

```
def color(x):
    """
    Color Conversion
    """
    x = tf.image.random_hue(x, 0.08)              # Adjust the hue of image
    x = tf.image.random_saturation(x, 0.6, 1.6)   # adjust image saturation
    x = tf.image.random_brightness(x, 0.05)       # adjust image brightness
    x = tf.image.random_contrast(x, 0.7, 1.3)     # adjust image contrast
    return x
image_2 = color(image_test)
image = np.hstack((image_test, image_2))
# Display image
plt.imshow(image)
```

Result:

- Image rotation

```
def rotate(x):
    """
    Rotate image
    """
    x = tf.image.rot90(x,tf.random.uniform(shape=[],minval=1,maxval=4,dtype=tf.int32))
    return x
image_2 = rotate(image_test)
image = np.hstack((image_test, image_2))
# Display image
plt.imshow(image)
```

Result:

- Zoom image

```python
def zoom(x, scale_min=0.6, scale_max=1.4):
    """

    Zoom Image
    """
    h, w, c = x.shape
    scale = tf.random.uniform([], scale_min, scale_max)
    sh = h * scale    # the height of image after zooming
    sw = w * scale    # the width of image after zooming
    x = tf.image.resize(x, (sh, sw))    # resize
    x = tf.image.resize_with_crop_or_pad(x, h, w)
    return x
image_2 = zoom(image_test)
image_2 = tf.cast(image_2, dtype=tf.uint8)
image = np.hstack((image_test, image_2))
# Display
plt.imshow(image)
```

Result:

- Setting data for training
  - Reload Dataset: because the dataset has been set before, it is necessary to reload

```python
train_data = tfds.load("cifar10", split=train_split)
```

- Converting data
  - Normalization: divide all pixels in the image by 255 to scale the pixel value between 0 and 1
  - Image augmentation: flip image horizontally, rotate image, convert colors, and zoom image
  - Label data: using one-hot encoding, for example, category 2 is represented as [0, 0, 0, 0, 0, 0, 0, 0, 1, 0]

```python
def parse_aug_fn(dataset):
    "
```

```
Image Augmentation function
"

x = tf.cast(dataset['image'], tf.float32) / 255.  # Image standardization

x = flip(x)  # Random horizontal flip

# color conversion

x = tf.cond(tf.random.uniform([], 0, 1) > 0.5, lambda: color(x), lambda: x)

# image rotation

x = tf.cond(tf.random.uniform([], 0, 1) > 0.75, lambda: rotate(x), lambda: x)

# image zoom

x = tf.cond(tf.random.uniform([], 0, 1) > 0.5, lambda: zoom(x), lambda: x)

# Convert the output label to One-hot encoding

y = tf.one_hot(dataset['label'], 10)

return x, y
```

- The training data after performing image augmentation

```
# shuffle data

train_data = train_data.shuffle(train_num)

# Loading data

train_data = train_data.map(map_func=parse_aug_fn, num_parallel_calls=AUTOTUNE)

# Set batch size and turn on prefetch mode

train_data = train_data.batch(batch_size).prefetch(buffer_size=AUTOTUNE)
```

- Displaying data after performing image augmentation

```
for images, labels in train_data.take(1):
    images = images.numpy()
# Create an array to display images
output = np.zeros((32 * 8, 32 * 8, 3))
# add 64 data into the array for displaying images
for i in range(8):
    for j in range(8):
        output[i*32:(i+1)*32, j*32:(j+1)*32, :] = images[i*8+j]
```

```
plt.figure(figsize=(8, 8))

# Display image

plt.imshow(output)
```

Result:

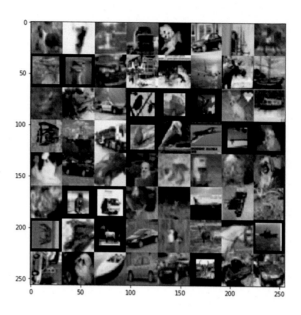

2. Building and training network models

Three multi-category classification models are implemented:
(1) Model-1: FCNN
(2) Model-2: CNN without applying data augmentation technique for training
(3) Model-3: apply data augmentation technique for training Model-2
Table 4.2 shows the architecture of Model-1, Model-2, and Model-3.

TABLE 4.2 The architecture of multi-category classification models.

Name	Architecture	Description
Model-1	- Input layer with shape of $32 \times 32 \times 3$ - Six fully connected layers, followed by ReLU activation functions - One dropout layer with discard rate of 30% - Output fully connected layer with 10 neurons, followed by softmax activation function	- FCNN - Do not apply data augmentation for training the model.
Model-2	- Input layer with shape of $32 \times 32 \times 3$ - Five convolutional layers, followed by ReLU activation function - One max pooling layer - One fully connected layer with 64 neurons, followed by ReLU activation function - Output fully connected layer with 10 neurons, followed by softmax activation function	- CNN - Do not apply data augmentation for training the model.
Model-3	- Input layer with shape of $32 \times 32 \times 3$ - Five convolutional layers, followed by ReLU activation function. - One max pooling layer - One fully connected layer with 64 neurons, followed by ReLU activation function - Output fully connected layer with 10 neurons, followed by softmax activation function	- CNN - Apply data augmentation for training the model.

(a) Model-1: FCNN
- Building network model

```
inputs = keras.Input(shape=(32, 32, 3))
x = layers.Flatten()(inputs)
x = layers.Dense(128, activation='relu')(x)
x = layers.Dense(256, activation='relu')(x)
x = layers.Dense(512, activation='relu')(x)
x = layers.Dense(512, activation='relu')(x)
x = layers.Dense(256, activation='relu')(x)
x = layers.Dense(64, activation='relu')(x)
x = layers.Dropout(0.3)(x)
outputs = layers.Dense(10, activation='softmax')(x)
# Create a network model
model_1 = keras.Model(inputs, outputs, name='model-1')
model_1.summary()    # Show network architecture
```

Result:

```
Model: "model-1"

Layer (type)                 Output Shape              Param #
=================================================================
input_2 (InputLayer)         [(None, 32, 32, 3)]       0

flatten_1 (Flatten)          (None, 3072)              0

dense_2 (Dense)              (None, 128)               393344

dense_3 (Dense)              (None, 256)               33024

dense_4 (Dense)              (None, 512)               131584

dense_5 (Dense)              (None, 512)               262656

dense_6 (Dense)              (None, 256)               131328

dense_7 (Dense)              (None, 64)                16448

dropout_1 (Dropout)          (None, 64)                0

dense_9 (Dense)              (None, 10)                650
=================================================================
Total params: 969,034
Trainable params: 969,034
Non-trainable params: 0
```

- Create a storage directory for saving model

```
model_dir = 'lab4-logs/models/'
os.makedirs(model_dir)
```

- Set callback function

```
# Save training records as TensorBoard log files
log_dir = os.path.join('lab4-logs', 'model-1')

model_cbk = keras.callbacks.TensorBoard(log_dir=log_dir)
# Save the best model
model_mckp = keras.callbacks.ModelCheckpoint(model_dir + '/Best-model-1.h5',
                        monitor='val_categorical_accuracy',
                        save_best_only=True,
                        mode='max')
```

- Set the optimizer, loss function, and metric function

```
model_1.compile(keras.optimizers.Adam(),
        loss=keras.losses.CategoricalCrossentropy(),
        metrics=[keras.metrics.CategoricalAccuracy()])
```

- Training Model-1

```
history_1 = model_1.fit(train_data,
            epochs=100,
            validation_data=valid_data,
            callbacks=[model_cbk, model_mckp])
```

Result:

```
Epoch 95/100
704/704 [==============================] - 9s 13ms/step - loss: 1.1343 - categorical_accuracy: 0.5854 - val_loss: 2.1138
 - val_categorical_accuracy: 0.4048
Epoch 96/100
704/704 [==============================] - 9s 13ms/step - loss: 1.1297 - categorical_accuracy: 0.5866 - val_loss: 2.0520
 - val_categorical_accuracy: 0.4068
Epoch 97/100

704/704 [==============================] - 9s 12ms/step - loss: 1.1219 - categorical_accuracy: 0.5902 - val_loss: 2.1084
 - val_categorical_accuracy: 0.4060
Epoch 98/100
704/704 [==============================] - 9s 12ms/step - loss: 1.1131 - categorical_accuracy: 0.5917 - val_loss: 2.1360
 - val_categorical_accuracy: 0.4086
Epoch 99/100
704/704 [==============================] - 9s 13ms/step - loss: 1.1165 - categorical_accuracy: 0.5928 - val_loss: 2.0842
 - val_categorical_accuracy: 0.3980
Epoch 100/100
704/704 [==============================] - 9s 13ms/step - loss: 1.1286 - categorical_accuracy: 0.5874 - val_loss: 2.0845
 - val_categorical_accuracy: 0.3904
```

(b) Model-2: CNN without applying data augmentation for training

- Building network model

```python
inputs = keras.Input(shape=(32, 32, 3))

x = layers.Conv2D(64, (3, 3), activation='relu')(inputs)

x = layers.MaxPool2D()(x)

x = layers.Conv2D(128, (3, 3), activation='relu')(x)

x = layers.Conv2D(256, (3, 3), activation='relu')(x)

x = layers.Conv2D(128, (3, 3), activation='relu')(x)

x = layers.Conv2D(64, (3, 3), activation='relu')(x)

x = layers.Flatten()(x)

x = layers.Dense(64, activation='relu')(x)

x = layers.Dropout(0.5)(x)

outputs = layers.Dense(10, activation='softmax')(x)


# Create model

model_2 = keras.Model(inputs, outputs, name='model-2')

model_2.summary()  # Show network architecture
```

Result:

```
Model: "model-2"

Layer (type)                    Output Shape            Param #
=================================================================
input_2 (InputLayer)            [(None, 32, 32, 3)]     0

conv2d (Conv2D)                 (None, 30, 30, 64)      1792

max_pooling2d (MaxPooling2D)    (None, 15, 15, 64)      0

conv2d_1 (Conv2D)               (None, 13, 13, 128)     73856

conv2d_2 (Conv2D)               (None, 11, 11, 256)     295168

conv2d_3 (Conv2D)               (None, 9, 9, 128)       295040

conv2d_4 (Conv2D)               (None, 7, 7, 64)        73792

flatten_1 (Flatten)             (None, 3136)            0

dense_7 (Dense)                 (None, 64)              200768

dropout (Dropout)               (None, 64)              0

dense_8 (Dense)                 (None, 10)              650
=================================================================
Total params: 941,066
Trainable params: 941,066
Non-trainable params: 0
```

- Set callback function

```python
# Save training records as TensorBoard log files
log_dir = os.path.join('lab4-logs', 'model-2')
model_cbk = keras.callbacks.TensorBoard(log_dir=log_dir)
# Save the best model
model_mckp = keras.callbacks.ModelCheckpoint(model_dir + '/Best-model-2.h5',
                    monitor='val_categorical_accuracy',
                    save_best_only=True,
                    mode='max')
```

- Set the optimizer, loss function, and metric function

```python
model_2.compile(keras.optimizers.Adam(),
        loss=keras.losses.CategoricalCrossentropy(),
        metrics=[keras.metrics.CategoricalAccuracy()])
```

- Training Model-2

```python
history_2 = model_2.fit(train_data,
            epochs=100,
            validation_data=valid_data,
            callbacks=[model_cbk, model_mckp])
```

Result：

```
Epoch 95/100
704/704 [==============================] - 14s 20ms/step - loss: 0.1105 - categorical_accuracy: 0.9664 - val_loss: 2.7155
 - val_categorical_accuracy: 0.7102
Epoch 96/100
704/704 [==============================] - 14s 20ms/step - loss: 0.1375 - categorical_accuracy: 0.9604 - val_loss: 2.5822
 - val_categorical_accuracy: 0.6940
Epoch 97/100
704/704 [==============================] - 15s 21ms/step - loss: 0.1115 - categorical_accuracy: 0.9663 - val_loss: 2.8166
 - val_categorical_accuracy: 0.6986
Epoch 98/100
704/704 [==============================] - 18s 26ms/step - loss: 0.1080 - categorical_accuracy: 0.9665 - val_loss: 2.8064
 - val_categorical_accuracy: 0.6910
Epoch 99/100
704/704 [==============================] - 15s 21ms/step - loss: 0.1243 - categorical_accuracy: 0.9630 - val_loss: 2.7873
 - val_categorical_accuracy: 0.6956
Epoch 100/100
704/704 [==============================] - 15s 21ms/step - loss: 0.1188 - categorical_accuracy: 0.9635 - val_loss: 2.9477
 - val_categorical_accuracy: 0.6898
```

(c) Model-3: CNN with data augmentation for training

```
inputs = keras.Input(shape=(32, 32, 3))
x = layers.Conv2D(64, (3, 3), activation='relu')(inputs)
x = layers.MaxPool2D()(x)
x = layers.Conv2D(128, (3, 3), activation='relu')(x)
x = layers.Conv2D(256, (3, 3), activation='relu')(x)
x = layers.Conv2D(128, (3, 3), activation='relu')(x)
x = layers.Conv2D(64, (3, 3), activation='relu')(x)
x = layers.Flatten()(x)
x = layers.Dense(64, activation='relu')(x)
x = layers.Dropout(0.5)(x)
outputs = layers.Dense(10, activation='softmax')(x)
# Create model
model_3 = keras.Model(inputs, outputs, name='model-3')
# show network architecuture
model_3.summary()
```

Result:

```
Model: "model-3"

_____
Layer (type)                 Output Shape              Param #
=================================================================
input_4 (InputLayer)         [(None, 32, 32, 3)]       0

conv2d_10 (Conv2D)           (None, 30, 30, 64)        1792

max_pooling2d_2 (MaxPooling2 (None, 15, 15, 64)        0

conv2d_11 (Conv2D)           (None, 13, 13, 128)       73856

conv2d_12 (Conv2D)           (None, 11, 11, 256)       295168

conv2d_13 (Conv2D)           (None, 9, 9, 128)         295040

conv2d_14 (Conv2D)           (None, 7, 7, 64)          73792

flatten_3 (Flatten)          (None, 3136)              0

dense_12 (Dense)             (None, 64)                200768

dropout_3 (Dropout)          (None, 64)                0

dense_13 (Dense)             (None, 10)                650
=================================================================
Total params: 941,066
Trainable params: 941,066
Non-trainable params: 0
_____
```

- Set callback function

```
## Save training records as TensorBoard log files
log_dir = os.path.join('lab4-logs', 'model-3')
model_cbk = keras.callbacks.TensorBoard(log_dir=log_dir)
```

```
# Save the best model

model_mckp = keras.callbacks.ModelCheckpoint(model_dir + '/Best-model-3.h5',

                    monitor='val_categorical_accuracy',

                    save_best_only=True,

                    mode='max')
```

- Set the optimizer, loss function, and metric function

```
model_3.compile(keras.optimizers.Adam(),

        loss=keras.losses.CategoricalCrossentropy(),

        metrics=[keras.metrics.CategoricalAccuracy()])
```

- Training Model-3

```
history_3 = model_3.fit(train_data,

            epochs=100,

            validation_data=valid_data,

            callbacks=[model_cbk, model_mckp])
```

Result:

```
704/704 [==============================] - 13s 19ms/step - loss: 0.8218 - categorical_accuracy: 0.7270 - val_loss: 0.6485
- val_categorical_accuracy: 0.7930
Epoch 96/100
704/704 [==============================] - 13s 19ms/step - loss: 0.8061 - categorical_accuracy: 0.7303 - val_loss: 0.6787
- val_categorical_accuracy: 0.7776
Epoch 97/100
704/704 [==============================] - 14s 19ms/step - loss: 0.8112 - categorical_accuracy: 0.7307 - val_loss: 0.6559
- val_categorical_accuracy: 0.7860
Epoch 98/100
704/704 [==============================] - 13s 19ms/step - loss: 0.8126 - categorical_accuracy: 0.7310 - val_loss: 0.6569
- val_categorical_accuracy: 0.7852
Epoch 99/100
704/704 [==============================] - 15s 21ms/step - loss: 0.8048 - categorical_accuracy: 0.7290 - val_loss: 0.6193
- val_categorical_accuracy: 0.8024
Epoch 100/100
704/704 [==============================] - 15s 22ms/step - loss: 0.8059 - categorical_accuracy: 0.7328 - val_loss: 0.6620
- val_categorical_accuracy: 0.7866
```

3. Displaying results
   - Load model weights

```
model_1.load_weights('lab4-logs/models/Best-model-1.h5')

model_2.load_weights('lab4-logs/models/Best-model-2.h5')

model_3.load_weights('lab4-logs/models/Best-model-3.h5')
```

- Verification on the test data

```
loss_1, acc_1 = model_1.evaluate(test_data)

loss_2, acc_2 = model_2.evaluate(test_data)

loss_3, acc_3 = model_3.evaluate(test_data)
```

- Display the loss value and accuracy result

```
loss = [loss_1, loss_2, loss_3]
acc = [acc_1, acc_2, acc_3]
dict = {"Accuracy": acc, "Loss": loss}
pd.DataFrame(dict)
```

Result:

	Loss	Accuracy
0	1.662437	0.4453
1	2.209503	0.7211
2	0.631140	0.7980

The results show that the CNN with image augmentation (Model-3) achieved the best accuracy, followed by the CNN alone (Model-2), and the FCNN (Model-3).

- Open TensorBoard: using command line to view training records

```
tensorboard --logdir lab4-logs
```

Fig. 4.25 shows the historical curve of three models on training data:

FIG. 4.25 Historical curve of three models on training data.

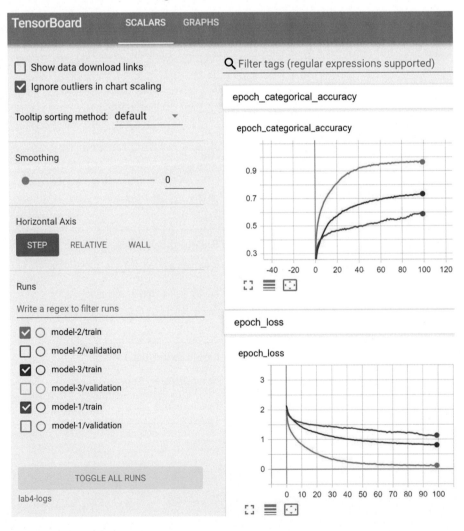

- Model-1 with pink line is the FCNN
- Model-2 with orange line is the CNN
- Model-3 with dark red line is the CNN using the image augmentation technique

Fig. 4.25 shows that Model-2 obtained the best result on the training set, followed by Model-3 and Model-1. Fig. 4.26 shows the historical curve of the three models on validation data:

- Model-1 with green line is the FCNN
- Model-2 with blue line is the CNN
- Model-3 with light cyan line is the CNN using the image augmentation technique.

Fig. 4.26 shows that Model-3 obtained the best results on the validation data, followed by Model-2 and Model-1. It also shows that Model-2 has an overfitting problem. When applying image augmentation for training, the overfitting problem was prevented and improved the accuracy of the Model-3.

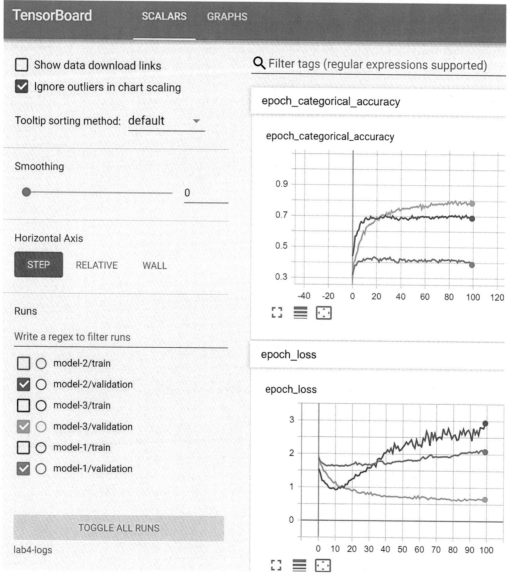

FIG. 4.26 Historical curve of three models on validation data.

References

[1] M.V. Valueva, N.N. Nagornov, P.A. Lyakhov, G.V. Valuev, N.I. Chervyakov, Application of the residue number system to reduce hardware costs of the convolutional neural network implementation, Math. Comput. Simul. 177 (2020) 232–243, https://doi.org/10.1016/j.matcom.2020.04.031. [Online]. Available.

[2] O. Russakovsky, J. Deng, H. Su, et al., ImageNet large scale visual recognition challenge, Int. J. Comput. Vis. 115 (2015) 211–252.

[3] C. Szegedy, W. Liu, Y. Jia, P. Sermanet, S. Reed, D. Anguelov, D. Erhan, V. Vanhoucke, A. Rabinovich, Going deeper with convolutions, in: Proceedings of the IEEE Conference on Computer Vision and Pattern Recognition, 2015, pp. 1–9.

[4] K. He, X. Zhang, S. Ren, J. Sun, Deep residual learning for image recognition, in: Proceedings of the IEEE Conference on Computer Vision and Pattern Recognition, 2016, pp. 770–778.

[5] S. Xie, R. Girshick, P. Dollár, T. Zhuowen, K. He, Aggregated residual transformations for deep neural networks, in: Proceedings of the IEEE Conference on Computer Vision and Pattern Recognition, 2017, pp. 1492–1500.

[6] K. Simonyan, A. Zisserman, Very deep convolutional networks for large-scale image recognition, in: International Conference on Learning Representations, 2015, pp. 1–14.

[7] M.D. Zeiler, R. Fergus, Visualizing and understanding convolutional networks, in: European Conference on Computer Vision, 2014, pp. 818–833.

[8] X. Ke, J. Zou, Y. Niu, End-to-end automatic image annotation based on deep CNN and multi-label data augmentation, IEEE Trans. Multimedia 21 (8) (2019) 2093–2106.

[9] J. Ding, B. Chen, H. Liu, M. Huang, Convolutional neural network with data augmentation for SAR target recognition, IEEE Geosci. Remote Sens. Lett. 13 (3) (2016) 364–368.

[10] X. Cui, V. Goel, B. Kingsbury, Data augmentation for deep neural network acoustic modeling, IEEE/ACM Trans. Audio Speech Lang. Process. 23 (9) (2015) 1469–1477.

[11] J. Salamon, J.P. Bello, Deep convolutional neural networks and data augmentation for environmental sound classification, IEEE Signal Process. Lett. 24 (3) (2017) 279–283.

[12] R. Dellana, K. Roy, Data augmentation in CNN-based periocular authentication, in: 2016 6th International Conference on Information Communication and Management (ICICM), Hatfield, 2016, pp. 141–145.

[13] Q. Hoang, T. Le, S. Huang, Data augmentation for improving SSD performance in rainy weather conditions, in: 2020 IEEE International Conference on Consumer Electronics - Taiwan (ICCE-Taiwan), Taoyuan, 2020, pp. 1–2.

[14] A. Krizhevsky, V. Nair, G. Hinton, Cifar-10 (Canadian institute for advanced research), [Online]. Available: http://www.cs.toronto.edu/~kriz/cifar.html.

5

Training neural network

5.1 Backpropagation

5.1.1 Introduction to backpropagation

Backpropagation (BP) [1] is a method to update the weights of neural networks in combination with gradient descent, which we presented in Chapter 2. Gradient descent is one of the most common algorithms to perform optimization for neural networks. By calculating the gradient and moving in the reverse direction, the error between the predicted result of the network and the expected output is reduced during training process. The weights update formula is:

$$W = W - \eta \frac{\partial L}{\partial W}$$

where L is loss function, η is learning rate, and W is weights of the neural network.

Through the preceding formula, the BP method is applied to compute the gradient of the loss function in respect to the weights of the neural network for each input and output example. The following is an example to explain how the gradient $\frac{\partial L}{\partial W}$ is calculated.

Fig. 5.1 shows the schematic diagram of a single-layer neural network with the forward propagation calculation. The computation process is:

$$z = x_1 w_1 + x_2 w_2 + x_3 w_3$$

$$\hat{y} = f(z)$$

$$Loss = (y - \hat{y})^2$$

where x_1, x_2, and x_3 are input data; w_1, w_2, and w_3 are weights of the network; f is an activation function; $\hat{y}$ is predicted value of the network, and y is an expected output.

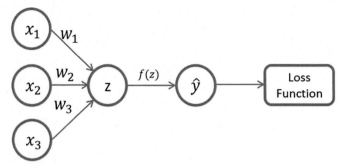

FIG. 5.1   Forward propagation calculation diagram of a single-layer neural network.

BP is applied to compute the gradients $\frac{\partial L}{\partial w_1}$, $\frac{\partial L}{\partial w_2}$, and $\frac{\partial L}{\partial w_3}$ for updating the weights of the network, as shown in Fig. 5.2. The computation process is:

$$\frac{\partial L}{\partial w_1} = \frac{\partial L}{\partial \hat{y}} \frac{\partial \hat{y}}{\partial z} \frac{\partial z}{\partial w1} = -2(y - \hat{y}) \times f'(z) \times x_1$$

$$\frac{\partial L}{\partial w_2} = \frac{\partial L}{\partial \hat{y}} \frac{\partial \hat{y}}{\partial z} \frac{\partial z}{\partial w_2} = -2(y - \hat{y}) \times f'(z) \times x_2$$

$$\frac{\partial L}{\partial w_3} = \frac{\partial L}{\partial \hat{y}} \frac{\partial \hat{y}}{\partial z} \frac{\partial z}{\partial w_3} = -2(y - \hat{y}) \times f'(z) \times x_3$$

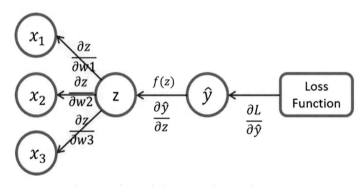

FIG. 5.2   Backpropagation (BP) computation diagram of a single-layer neural network.

5.1.2 Vanishing gradient problem

The vanishing gradient refers to the problem encountered when training neural networks, especially deep neural networks (DNNs) with gradient methods and BP.

Table 5.1 lists three common activation functions: sigmoid, tanh, and rectified linear unit (ReLU). The use of sigmoid and tanh activation functions in hidden layers of the neural network may cause the problem of vanishing gradient. For example, when performing BP through a layer of sigmoid activation function, it needs to be multiplied with gradient of sigmoid function. The maximum value of the sigmoid derivative function is $f'(0) = 0.25$, which means that the gradient will be attenuated 0.25 times after one pass. If the neural network has five layers, the gradient will be reduced by at least 0.25^5, which makes it difficult to update the weights of the previous layers of the neural network. Similarly, tanh activation function has the same situation. Recently, the ReLU activation function is usually employed in hidden layers because it outputs the input directly if the input is positive, otherwise, it outputs zero, overcoming the vanishing gradient problem and allowing the neural networks to learn faster and more effectively.

TABLE 5.1 Three common activation functions used in neural networks.

	$f(z)$	$f'(z)$	Plot
Sigmoid	$\frac{1}{1+e^{-z}}$	$f(z)(1-f(z))$	
Tanh	$\frac{e^{z}-e^{-z}}{e^{z}+e^{-z}}$	$1-f(z)^2$	
ReLU	$\begin{cases} 0 \, for \, x<0 \\ x \, for \, x \geq 0 \end{cases}$	$\begin{cases} 0 \, for \, x<0 \\ 1 \, for \, x \geq 0 \end{cases}$	

Here we explain vanishing gradient when using sigmoid activation function in neural networks. Given a two-layer neural network, in which each layer uses the sigmoid activation function (f) to compute an output value, input $x = 0.4$, weights $w_1 = 1$, and $w_2 = 1$, and expected output $y = 1$. Fig. 5.3 shows a schematic diagram of the network with forward propagation computation. The computation process is:

$$h_1 = xw_1 \cong 0.4$$

$$h_2 = f(h_1) \cong 0.599$$
$$h_3 = h_2 w_2 \cong 0.599$$
$$\hat{y} = f(h_3) \cong 0.645$$
$$Loss = (y - \hat{y})^2 \cong 0.126$$

here

$$f(x) = \frac{1}{1 + e^{-x}}$$

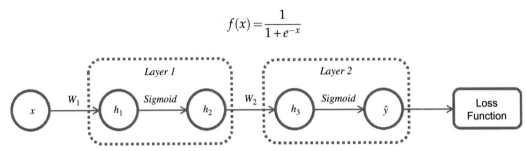

FIG. 5.3   Forward propagation computation diagram of a two-layer neural network.

As shown in Fig. 5.4, $\frac{\partial L}{\partial w_1}$ is calculated by the BP method, the computation process is:

$$\frac{\partial L}{\partial w_1} = \frac{\partial L}{\partial \hat{y}} \frac{\partial \hat{y}}{\partial h_3} \frac{\partial h_3}{\partial h_2} \frac{\partial h_2}{\partial h_1} \frac{\partial h_1}{\partial w_1} = -2(y - \hat{y}) \times f'(h_3) \times w_2 \times f'(h_1) \times x =$$
$$-2(1 - 0.645) \times f'(0.599) \times 1 \times f'(0.4) \times 0.4 =$$
$$-2(1 - 0.645) \times 0.24 \times 1 \times 0.229 \times 0.4 = -0.0156$$

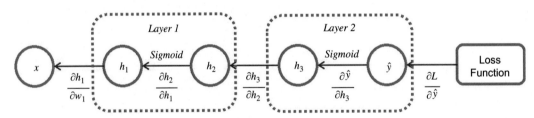

FIG. 5.4   Backpropagation (BP) computation diagram of a two-layer neural network.

From the computation of the BP above, it can be found that after each sigmoid activation function, the computed value will be reduced at least 0.25 times, so the more layers of transfer, the more gradient attenuation there will be. This is the reason for vanishing gradient during training DNNs.

The computation result of $\frac{\partial L}{\partial w_1}$ is verified by using the TensorFlow program, as shown here.

```python
import tensorflow as tf
# Declare input, weight and expected output
x = 0.4
w1, w2 = tf.Variable(1.0), tf.Variable(1.0)
y = 1
# Forward propagation will be recorded in "tape"
with tf.GradientTape() as tape:
    # Forward calculation
    h1 = x * w1
    h2 = tf.sigmoid(h1)
    h3 = h2 * w2
    y_hat = tf.sigmoid(h3)
    loss = (y - y_hat)**2
# backpropagation "tape" to calculate the weight w1
gradients = tape.gradient(loss, w1)
print(gradients)
```

Result: tf.Tensor(-0.015601176, shape=(), dtype=float32)

5.2 Weight initialization

The weight initialization plays an important role in training neural networks because it directly affects the convergence of the model [2–5]. In this section, we introduce and analyze three weight initialization methods and the impact of them on building the neural networks. In the following, we give a brief description of the three different weight initialization methods.

- Normal distribution initialization: The simplest case of a normal distribution is known as the standard normal distribution when the parameter μ (mean) is set to 0 and the parameter σ (standard deviation) is set to 1. In the experiment below, the neural network of sigmoid activation function is used to analyze the initial weights of the normal distribution with μ = 0 and σ = 1, and the normal distribution with μ = 0 and σ = 0.01. The final analysis results show that the normal distribution with μ = 0 and σ = 1 is a cause of the vanishing gradient.
- Xavier or Glorot initialization [6]: This is used to improve the problems of the normal distribution initialization method by trying to keep the scale of the gradient roughly the same in all layers of the neural network. However, using Glorot initialization for DNNs with the ReLU activation function still reveals the vanishing gradient during the training process.
- He initialization [7]: This addresses the problem of Xavier or Glorot initialization when the ReLU activation function is employed in DNNs by making the output distribution of each layer even.

The following introduces in detail and analyzes the impacts of each initialization method in training and developing neural networks. We also list the necessary packages for the program examples.

```python
import numpy as np
import tensorflow as tf
import matplotlib.pyplot as plt
from tensorflow import keras
from tensorflow.keras import layers
from tensorflow.keras import initializers
```

5.2.1 Normal Distribution

1. The neural network of sigmoid activation function and the normal distribution with μ = 0 and σ = 1

To build the network model, the following network layers are used:

- keras.Input: input layer with shape of (100,).
- layers.Dense (fully connected layer): sigmoid is used as the activation function, and normal distribution with μ = 0 and σ = 1 is employed for weight initialization; bias is not used.

Building the network:

```
inputs = keras.Input(shape=(100,))
x1 = layers.Dense(100, 'sigmoid', False, initializers.RandomNormal(0, 1))(inputs)
x2 = layers.Dense(100, 'sigmoid', False, initializers.RandomNormal(0, 1))(x1)
x3 = layers.Dense(100, 'sigmoid', False, initializers.RandomNormal(0, 1))(x2)
x4 = layers.Dense(100, 'sigmoid', False, initializers.RandomNormal(0, 1))(x3)
x5 = layers.Dense(100, 'sigmoid', False, initializers.RandomNormal(0, 1))(x4)
model_1 = keras.Model(inputs, [x1, x2, x3, x4, x5])
```

Display the output distribution of each layer:

```
x = np.random.randn(100, 100)
outputs = model_1.predict(x)
for i, layer_output in enumerate(outputs):
    plt.subplot(1, 5, i+1)  # Choose which cell to display in the table
    plt.title(str(i+1) + "-layer")  # Set the title of the histogram
    if i != 0: plt.yticks([], [])  # Show only the y-axis of the first column of histograms
    plt.hist(layer_output.flatten(), 30, range=[0,1])  # Draw a histogram
plt.show()
```

Result:

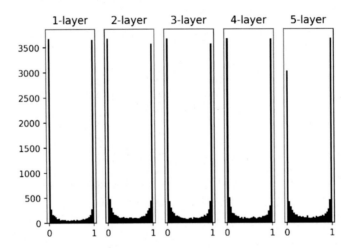

The results show that most of the output values are distributed near 0 and 1. The gradient values will approach zero when applying BP (please refer to the derivative of sigmoid function in Table 5.1), resulting in the vanishing gradient problem. Therefore, the normal distribution with $\mu = 0$ and $\sigma = 1$ is not an effective weight initialization method.

2. The neural network of sigmoid activation function and the normal distribution with $\mu = 0$ and $\sigma = 0.01$

To build the network model, the following network layers are used:

- keras.Input: input layer with shape of (100,)
- layers.Dense (fully connected layer): sigmoid is used as activation function, and normal distribution with $\mu = 0$ and $\sigma = 0.01$ is employed for weight initialization; bias is not used

Building the network:

```
inputs = keras.Input(shape=(100,))
x1 = layers.Dense(100, 'sigmoid', False, initializers.RandomNormal(0, 0.01))(inputs)
x2 = layers.Dense(100, 'sigmoid', False, initializers.RandomNormal(0, 0.01))(x1)
x3 = layers.Dense(100, 'sigmoid', False, initializers.RandomNormal(0, 0.01))(x2)
x4 = layers.Dense(100, 'sigmoid', False, initializers.RandomNormal(0, 0.01))(x3)
x5 = layers.Dense(100, 'sigmoid', False, initializers.RandomNormal(0, 0.01))(x4)
model_2 = keras.Model(inputs, [x1, x2, x3, x4, x5])
```

Display the output distribution of each layer:

```
x = np.random.randn(100, 100)
outputs = model_2.predict(x)
for i, layer_output in enumerate(outputs):
    plt.subplot(1, 5, i+1)  # Choose which cell to display in the table
    plt.title(str(i+1) + "-layer")  # Set the title of the histogram
    if i != 0: plt.yticks([], [])  # Show only the y-axis of the first column of histograms
    plt.hist(layer_output.flatten(), 30, range=[0,1])  # Draw a histogram
plt.show()
```

Result:

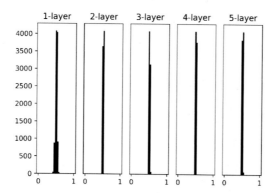

The results show that the output values from the second layer are mostly distributed around 0.5, and the gradient values will be around 0.25 when applying BP (please refer to the derivative of sigmoid function in Table 5.1). Although the problem of vanishing gradient is improved, it may still occur with a deep network that consists of many convolutional layers.

5.2.2 Glorot initialization

1. The neural network with Glorot initialization and sigmoid activation function

Glorot et al. [6] proposed the Glorot initialization method in 2010. It has been widely used in many DNNs and is regarded as the default weight initialization method for training network models in Keras.

To build the network model, the following network layers are used:

- keras.Input: input layer with shape of (100,)
- layers.Dense (fully connected layer): sigmoid is used as activation function, and the Glorot method is employed for weight initialization; bias is not used

Building the network:

```
inputs = keras.Input(shape=(100,))
x1 = layers.Dense(100, 'sigmoid', False, initializers.glorot_normal())(inputs)
x2 = layers.Dense(100, 'sigmoid', False, initializers.glorot_normal())(x1)
```

```
x3 = layers.Dense(100, 'sigmoid', False, initializers.glorot_normal())(x2)
x4 = layers.Dense(100, 'sigmoid', False, initializers.glorot_normal())(x3)
x5 = layers.Dense(100, 'sigmoid', False, initializers.glorot_normal())(x4)
model_3 = keras.Model(inputs, [x1, x2, x3, x4, x5])
```

Display the output distribution of each layer:

```
x = np.random.randn(100, 100)
outputs = model_3.predict(x)
for i, layer_output in enumerate(outputs):
    plt.subplot(1, 5, i+1)  # Choose which cell to display in the table
    plt.title(str(i+1) + "-layer")  # Set the title of the histogram
    if i != 0: plt.yticks([], [])  # Show only the y-axis of the first column of histograms
    plt.hist(layer_output.flatten(), 30, range=[0,1])  # Draw a histogram
plt.show()
```

Result:

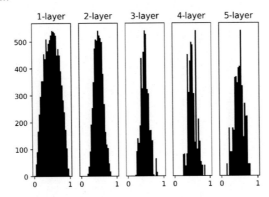

The results show that the Glorot initialization method produces a wider output distribution than that of the normal distribution method. This helps to prevent the vanishing gradient during the training process and allows the neural networks to learn data efficiently. Therefore, the Glorot method for weights initialization is recommended for neural networks with sigmoid or tanh activation function.

2. The neural network with Glorot initialization and ReLU activation function

To build the network model, the following layers are used:

- keras.Input: input layer with shape of (100,)
- layers.Dense (fully connected layer): ReLU is used as the activation function, and Glorot is employed for weight initialization; bias is not used.

Building the network

```
inputs = keras.Input(shape=(100,))
x1 = layers.Dense(100, 'relu', False, initializers.glorot_normal())(inputs)
x2 = layers.Dense(100, 'relu', False, initializers.glorot_normal())(x1)
x3 = layers.Dense(100, 'relu', False, initializers.glorot_normal())(x2)
x4 = layers.Dense(100, 'relu', False, initializers.glorot_normal())(x3)
x5 = layers.Dense(100, 'relu', False, initializers.glorot_normal())(x4)
model_4 = keras.Model(inputs, [x1, x2, x3, x4, x5])
```

Display the output distribution of each layer:

```
x = np.random.randn(100, 100)
outputs = model_4.predict(x)
for i, layer_output in enumerate(outputs):
    plt.subplot(1, 5, i+1)  # Choose which cell to display in the table
    plt.title(str(i+1) + "-layer")  # Set the title of the histogram
    if i != 0: plt.yticks([], [])  # Show only the y-axis of the first column of histograms
    plt.hist(layer_output.flatten(), 30, range=[0,1])  # Draw a histogram
plt.show()
```

Result:

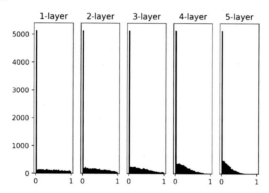

The results show that when the number of layers of the network model increases, the output distribution of the deeper layer is closer to 0, which may cause the vanishing gradient problem during training process. Therefore, using Glorot weight initialization method for DNNs of the ReLU activation function is not recommend

5.2.3 He initialization

1. The neural network with He initialization and ReLU activation function

Kaiming He proposed the He initialization method [7] in 2015. It has proven effective in preventing the vanishing gradient problem when using DNNs with the ReLU activation function.

To build the network model, the following layers are used:

- keras.Input: input layer with shape of (100,)
- layers.Dense (fully connected layer): ReLU is used as activation function, and He method is employed for weight initialization; bias is not used

Building the network:

```
inputs = keras.Input(shape=(100,))
x1 = layers.Dense(100, 'relu', False, initializers.he_normal())(inputs)
x2 = layers.Dense(100, 'relu', False, initializers.he_normal())(x1)
x3 = layers.Dense(100, 'relu', False, initializers.he_normal())(x2)
x4 = layers.Dense(100, 'relu', False, initializers.he_normal())(x3)
x5 = layers.Dense(100, 'relu', False, initializers.he_normal())(x4)
model_5 = keras.Model(inputs, [x1, x2, x3, x4, x5])
```

Display the output distribution of each layer:

```
x = np.random.randn(100, 100)
outputs = model_5.predict(x)
for i, layer_output in enumerate(outputs):
    plt.subplot(1, 5, i+1)  # Choose which cell to display in the table
    plt.title(str(i+1) + "-layer")  # Set the title of the histogram
    if i != 0: plt.yticks([], [])  # Show only the y-axis of the first column of histograms
    plt.hist(layer_output.flatten(), 30, range=[0,1])  # Draw a histogram
plt.show()
```

Result:

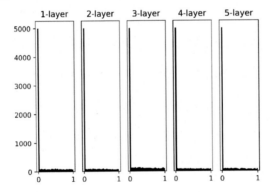

The results show that the output distribution of each layer is very even when a deep network with ReLU activation uses the He initialization method. The problem of using Glorot initialization method for DNNs with ReLU activation function is successfully solved.

The following points summarize the knowledge of this section.

- The Glorot method is recommended for weight initialization when using neural networks with sigmoid or tanh activation functions.
- The He method is recommended for weight initialization when using neural networks with ReLU activation function.
- DNNs with the ReLU activation function usually achieve better performance than that of DNNs with sigmoid or tanh activation functions.

5.3 Batch normalization

5.3.1 Introduction to batch normalization

Ioffe and Szegedy [8] proposed the batch normalization algorithm in 2015. It allows the use of high learning rates for speeding up training of neural networks while placing less emphasis on weight initialization. When training neural networks, the output distribution of the network layers is changed. The latter layer must continuously adapt to the output distribution changes of the previous layer, that is, each layer needs to be adjusted according to the output distribution of the previous layer. This makes the neural networks update the weights slowly because of requiring careful weight initialization and slow learning rates. To address this problem, the core idea of batch normalization is to perform the normalization for each batch of input data and output at each layer of the neural networks. As shown in Fig. 5.5A, the output of the network layer is evenly distributed between −10 and 10. After passing through the tanh activation function, the output distribution is not even; most output values are distributed near −1 and 1, thus causing the problem of the vanishing gradient (please refer to the derivative of tanh function in Table 5.1). In Fig. 5.5B, the output of the network layer is firstly normalized to the range of about −2 to 2, and then it is passed through the tanh activation function, resulting in more even output distribution compared to the case without normalization in Fig. 5.5A.

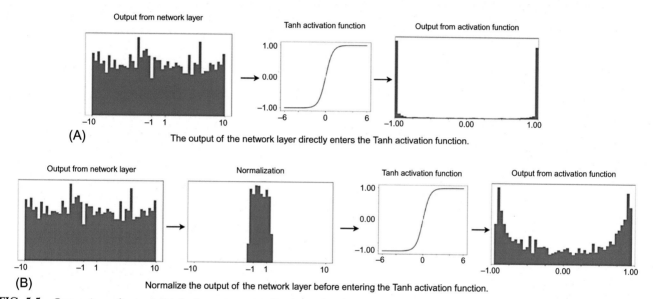

FIG. 5.5   Comparison of output distribution using normalization and without using normalization.

With input mini-batch data $M = M = \{x_1, x_2, \ldots x_n\}$, batch normalization transform is expressed through four steps:

- Step 1: computing the mean of mini-batch:

$$\mu_M = \frac{1}{n} \sum_{i=1}^{n} x_i$$

- Step 2: computing the variance of mini-batch:

$$\sigma_M^2 = \frac{1}{n} \sum_{i=1}^{n} (x_i - \mu_M)^2$$

- Step 3: normalizing the value:

$$\hat{x}_i = \frac{x_i - \mu_M}{\sqrt{\sigma_M^2 + \varepsilon}}$$

where ε is a constant for numerical stability.

- Step 4: scaling and shifting:

$$y_i = \alpha \hat{x}_i + \beta$$

where y_i represents the output of batch normalization transform, and α and β are parameters to be learned.

By adding the batch normalization transform to the neural networks, any activation can be manipulated. In the next section, we present the location of batch normalization in the neural network.

The following points summarize the advantages of batch normalization:

- It speeds up the training of neural networks by using higher learning rates
- It reduces the overfitting problem in training neural networks
- It reduces the vanishing gradient and exploding gradient problems
- It eliminates the need to use the dropout technique to prevent loss of data information

5.3.2 Neural network with batch normalization

Batch normalization has been applied successfully to neural networks to improve the performance of various computer vision applications such as image classification [9], image recognition [10], object detection [11], image-to-image translation [12], and so on. The neural networks with batch normalization are slightly different from the normal neural networks, in which the location of the batch normalization is between the convolutional layer and the activation function, as shown in Fig. 5.6.

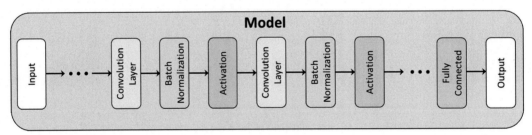

FIG. 5.6   The structure of a convolution neural network with batch normalization.

Here, we present a code example of adding batch normalization to neural networks.

- Original network construction:

```
......
x = layers.Conv2D(128, (3, 3), activation='relu')(x)
......
```

- Adding Batch Normalization to the layer:

```
......
x = layers.Conv2D(128, (3, 3))(x)
x = layers.BatchNormalization()(x)
x = layers.ReLU()(x)
......
```

5.4 Experiment 1: Verification of three weight initialization methods

This section continues the topic of CIFAR-10 image classification in Chapter 4. To verify the effectiveness of each weight initialization method, three scenarios of building neural networks are implemented:

(1) Model-1: a neural network with normal distribution ($\mu = 0$ and $\sigma = 0.01$)
(2) Model-2: a neural network with Glorot weight initialization
(3) Model-3: a neural network with He weight initialization

Model-1, Model-2, and Model-3 have the same architecture and use ReLU activation in the hidden layers.

5.4.1 Code examples

Fig. 5.7 shows the flowchart of the source code for building and testing three models on CIFAR-10 dataset.

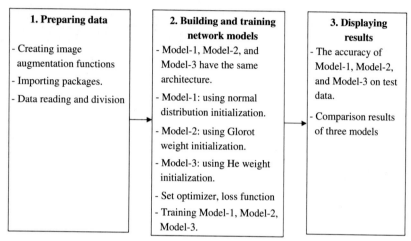

FIG. 5.7 The flowchart of the source code for CIFAR-10 image classification.

1. Preparing data
    (a) Creating image augmentation functions

Because the image augmentation technique is used in the next chapters of the book, a Python file of the image augmentation functions is created here. To use this technique, directly import its function to the program.

- Create file: create a file in Jupyter, as shown in Fig. 5.8

FIG. 5.8 Creating a file.

- Rename: change the file name to "preprocessing.py," as shown in Fig. 5.9

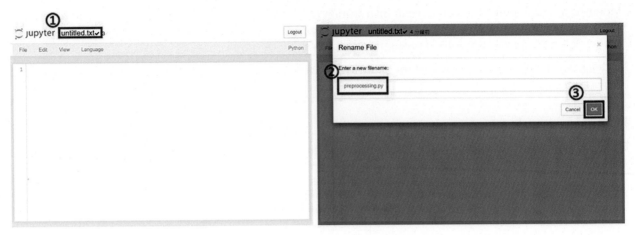

FIG. 5.9 Modify the file name.

- Import necessary packages

```
import tensorflow as tf
```

- Create image augmentation functions: write image augmentation functions to the "preprocessing.py" file.

```
def flip(x):
    "
    flip image
    "
    x = tf.image.random_flip_left_right(x)  # Random flip image, left and right
    return x

def color(x):
```

```python
"
Change Color
"
x = tf.image.random_hue(x, 0.08)  # Adjust the hue of image by a random factor
x = tf.image.random_saturation(x, 0.6, 1.6)  # Randomly adjust image saturation
x = tf.image.random_brightness(x, 0.05)  # Randomly adjust image brightness
x = tf.image.random_contrast(x, 0.7, 1.3)  # Randomly adjust image contrast
return x


def rotate(x):
    "
    Rotate image
    "
    # Randomly select n times (set the range of n through minval and maxval),
    x = tf.image.rot90(x,tf.random.uniform(shape=[],minval=1,maxval=4,dtype=tf.int32))
    return x


def zoom(x, scale_min=0.6, scale_max=1.4):
    "
    Zoom Image
    "
    h, w, c = x.shape
    scale = tf.random.uniform([], scale_min, scale_max)  # Random scaling
    sh = h * scale  # the height of image after zooming
    sw = w * scale  # the width of image after zooming
    x = tf.image.resize(x, (sh, sw))  # Image zoom
    x = tf.image.resize_with_crop_or_pad(x, h, w)  # resize image
    return x
```

- Data Preprocessing: Add data preprocessing functions in the "preprocessing.py" file.

```python
def parse_aug_fn(dataset):
    "
    Image Augmentation function
    "
    x = tf.cast(dataset['image'], tf.float32) / 255.  # Image standardization
    x = flip(x)  # Random horizontal flip
    # color conversion
    x = tf.cond(tf.random.uniform([], 0, 1) > 0.5, lambda: color(x), lambda: x)
```

```
    # image rotation
    x = tf.cond(tf.random.uniform([], 0, 1) > 0.75, lambda: rotate(x), lambda: x)
    # image zoom
    x = tf.cond(tf.random.uniform([], 0, 1) > 0.5, lambda: zoom(x), lambda: x)
    # Convert the output label to One-hot encoding
    y = tf.one_hot(dataset['label'], 10)
    return x, y


def parse_fn(dataset):
    x = tf.cast(dataset['image'], tf.float32) / 255.  # Image standardization
    # Convert the output label to One-hot encoding
    y = tf.one_hot(dataset['label'], 10)
    return x, y
```

b) Import packages

```
import os
import numpy as np
import pandas as pd
import tensorflow as tf
import tensorflow_datasets as tfds
import matplotlib.pyplot as plt
from tensorflow import keras
from tensorflow.keras import layers
from tensorflow.keras import initializers
# Import "parse_aug_fn" function and "parse_fn" function from the preprocessing.py file
from preprocessing import parse_aug_fn, parse_fn
```

c) Data reading and division

■ Load CIFAR-10 dataset:

```
# Divide the training data with the rate of 9: 1 (9 for training and 1 for validation)
train_split, valid_split = ['train[:90%]', 'train[90%:]']
# get the training data and read data information
train_data, info = tfds.load("cifar10", split=train_split, with_info=True)
# get the valid data
valid_data = tfds.load("cifar10", split=valid_split)
# get the test set of CIFAR-10
test_data = tfds.load("cifar10", split=tfds.Split.TEST)
```

■ Data settings

```
AUTOTUNE = tf.data.experimental.AUTOTUNE  # Automatic adjustment mode
batch_size = 64  # Batch size
train_num = int(info.splits['train'].num_examples / 10) * 9  # Number of training data


train_data = train_data.shuffle(train_num)  # Shuffle the training data
# Training data
train_data = train_data.map(map_func=parse_aug_fn, num_parallel_calls=AUTOTUNE)
# Set batch size and turn on prefetch mode
train_data = train_data.batch(batch_size).prefetch(buffer_size=AUTOTUNE)


# Validation data
valid_data = valid_data.map(map_func=parse_fn, num_parallel_calls=AUTOTUNE)
# Set batch size and turn on prefetch mode
valid_data = valid_data.batch(batch_size).prefetch(buffer_size=AUTOTUNE)


# Test data
test_data = test_data.map(map_func=parse_fn, num_parallel_calls=AUTOTUNE)
#Set batch size and turn on prefetch mode
test_data = test_data.batch(batch_size).prefetch(buffer_size=AUTOTUNE)
```

2. Building and training network models

Three CIFAR-10 image classification models with the same architecture are implemented. Table 5.2 lists the architecture and description of each model.

TABLE 5.2 The architecture of CIFAR-10 image classification models.

Name	Architecture	Description
Model-1	- Input layer with shape of (32,32,3) - Five convolutional layers, followed by ReLU activation function - One max pooling layer - One flatten layer for flattening the input into a one-dimensional Tensor - One fully connected layer - One dropout layer with a discard rate of 50% - Output fully connected layer with 10 neurons, followed by softmax function	Using normal distribution method with $\mu = 0$ and $\sigma = 0.01$ for weight initialization when training the model
Model-2	- Input layer with shape of (32, 32, 3) - Five convolutional layers, followed by ReLU activation function - One max pooling layer - One flatten layer for flattening the input into a one-dimensional Tensor - One fully connected layer - One dropout layer with a discard rate of 50% - Output fully connected layer with 10 neurons, followed by softmax function	Using Glorot method for weight initialization when training the model
Model-3	- Input layer with shape of (32, 32, 3) - Five convolutional layers, followed by ReLu activation function - One max pooling layer - One flatten layer for flattening the input into a one-dimensional Tensor - One fully connected layer, followed by ReLU activation function - One dropout layer with a discard rate of 50% - Output fully connected layer with 10 neurons, followed by softmax function	Using He method for weight initialization when training the model

- Building network models:

```
def build_and_train_model(run_name, init):
    "
    run_name: the name of the current executing task
    init: weight initialization method
    "
    inputs = keras.Input(shape=(32, 32, 3))
    x = layers.Conv2D(64, (3, 3), activation='relu', kernel_initializer=init)(inputs)
```

```
    x = layers.MaxPool2D()(x)
    x = layers.Conv2D(128, (3, 3), activation='relu', kernel_initializer=init)(x)
    x = layers.Conv2D(256, (3, 3), activation='relu', kernel_initializer=init)(x)
    x = layers.Conv2D(128, (3, 3), activation='relu', kernel_initializer=init)(x)
    x = layers.Conv2D(64, (3, 3), activation='relu', kernel_initializer=init)(x)
    x = layers.Flatten()(x)
    x = layers.Dense(64, activation='relu', kernel_initializer=init)(x)
    x = layers.Dropout(0.5)(x)
    outputs = layers.Dense(10, activation='softmax')(x)
    # Create a network model (connect all the network layers that pass through from input
to output)
    model = keras.Model(inputs, outputs)

    # Save training log
    logfiles = 'lab5-logs/{}-{}'.format(run_name, init.__class__.__name__)
    # save the weight distribution of each layer
    model_cbk = keras.callbacks.TensorBoard(log_dir=logfiles,
                            histogram_freq=1)
    # save the best weights of the model
    modelfiles = model_dir + '/{}-best-model.h5'.format(run_name)
    model_mckp = keras.callbacks.ModelCheckpoint(modelfiles,
                        monitor='val_categorical_accuracy',
                        save_best_only=True,
                        mode='max')
    # Set the optimizer, loss function, and metric function for training
    model.compile(keras.optimizers.Adam(),
            loss=keras.losses.CategoricalCrossentropy(),
            metrics=[keras.metrics.CategoricalAccuracy()])

    # Train the network model
    model.fit(train_data,
        epochs=100,
        validation_data=valid_data,
        callbacks=[model_cbk, model_mckp])
```

- Training the network model using three weight initialization methods:

```
session_num = 1
# Set storage weight directory
model_dir = 'lab5-logs/models/'
```

```
os.makedirs(model_dir)
# Set the three weight initialization methods
weights_initialization_list = [initializers.RandomNormal(0, 0.01),
                    initializers.glorot_normal(),
                    initializers.he_normal()]

for init in weights_initialization_list:
    print('--- Running training session %d' % (session_num))
    run_name = "run-%d" % session_num
    build_and_train_model(run_name, init) # Create and train a network
    session_num += 1
```

Result:

```
Epoch 95/100
704/704 [==============================] - 13s 19ms/step - loss: 0.7215 - categorical_accuracy: 0.7616 - val_loss: 0.6090
- val_categorical_accuracy: 0.8026
Epoch 96/100
704/704 [==============================] - 13s 19ms/step - loss: 0.7239 - categorical_accuracy: 0.7618 - val_loss: 0.5869
- val_categorical_accuracy: 0.8136
Epoch 97/100
704/704 [==============================] - 13s 19ms/step - loss: 0.7152 - categorical_accuracy: 0.7641 - val_loss: 0.6050
- val_categorical_accuracy: 0.8096
Epoch 98/100
704/704 [==============================] - 13s 19ms/step - loss: 0.7254 - categorical_accuracy: 0.7603 - val_loss: 0.6007
- val_categorical_accuracy: 0.8082
Epoch 99/100
704/704 [==============================] - 14s 19ms/step - loss: 0.7207 - categorical_accuracy: 0.7644 - val_loss: 0.5784
- val_categorical_accuracy: 0.8144
Epoch 100/100
704/704 [==============================] - 14s 19ms/step - loss: 0.7297 - categorical_accuracy: 0.7622 - val_loss: 0.6065
- val_categorical_accuracy: 0.8162
```

3. Displaying results

- Load the best trained models:

```
model_1 = keras.models.load_model('lab5-logs/models/run-1-best-model.h5')
model_2 = keras.models.load_model('lab5-logs/models/run-2-best-model.h5')
model_3 = keras.models.load_model('lab5-logs/models/run-3-best-model.h5')
```

- Verification on the test set:

```
loss_1, acc_1 = model_1.evaluate(test_data)
loss_2, acc_2 = model_2.evaluate(test_data)
loss_3, acc_3 = model_3.evaluate(test_data)
```

- Display the loss value and accuracy result：

```
loss = [loss_1, loss_2, loss_3]
acc = [acc_1, acc_2, acc_3]
dict = {"Loss": loss, "Accuracy": acc}
pd.DataFrame(dict)
```

Result:

	Loss	Accuracy
0	2.302650	0.1000
1	0.635572	0.7933
2	0.609081	0.8118

The results show that Model-3 with ReLU activation function and He weight initialization achieved the greatest accuracy, followed by the Model-2 with Glorot weight initialization and Model-3 with normal distribution ($\mu = 0$ and $\sigma = 0.01$).

5.4.2 Visualizing weight distribution with TensorBoard

TensorBoard provides two kinds of data distribution visualization tools; DISTRIBUTIONS and HISTOGRAMS. When training the network, the tf.keras.callbacks.TensorBoard function has the "histogram_freq" parameter set to 1, which means that the weight distribution of each layer in each epoch training is recorded, and both DISTRIBUTIONS and HISTOGRAMS can be used.

The following uses directly the training results of three models in the previous section with three different weight initialization methods including He initialization, Glorot initialization, and normal distribution ($\mu = 0$ and $\sigma = 0.01$) to visualize and analyze the weight distribution changes through TensorBoard.

Go to the location where the "lab5-logs" is stored, then open TensorBoard through the Command line below to view training records.

```
tensorboard --logdir lab5-logs
```

1. The weights distribution of the Model-1 with normal distribution initialization
   (a) Using the DISTRIBUTIONS tool

Fig. 5.10 shows the weight changes of Model-1 in 100 epochs, where the x-axis is the time axis and the y-axis is the range of weight distribution. It can be observed that the weight distribution of the first and second convolutional layers has changed very little. It is obvious that the network weights have not been updated at all.

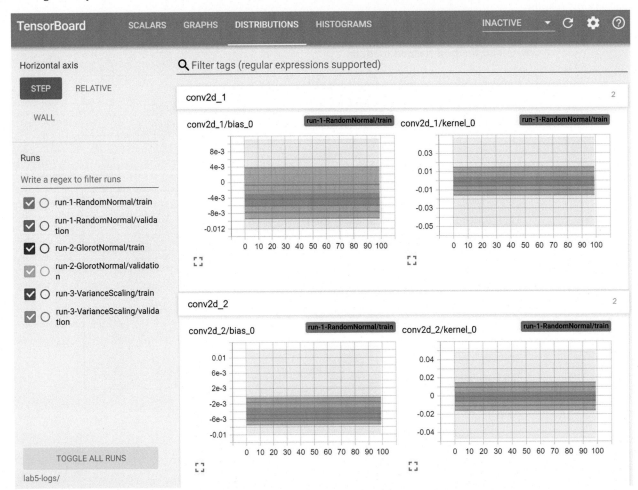

FIG. 5.10   DISTRIBUTIONS shows the weight changes of Model-1 with normal distribution ($\mu = 0$ and $\sigma = 0.01$). The top left and top right of the figure are the bias distribution and kernel distribution of the second convolution layer, respectively. The bottom left and bottom right of the figure are the bias distribution and kernel distribution of the third convolution layer, respectively.

(b) Using HISTOGRAMS tool

Fig. 5.11 shows another form of weight changes in 100 epochs of Model-1, where the x-axis is the range of weight distribution and the y-axis is the time axis.

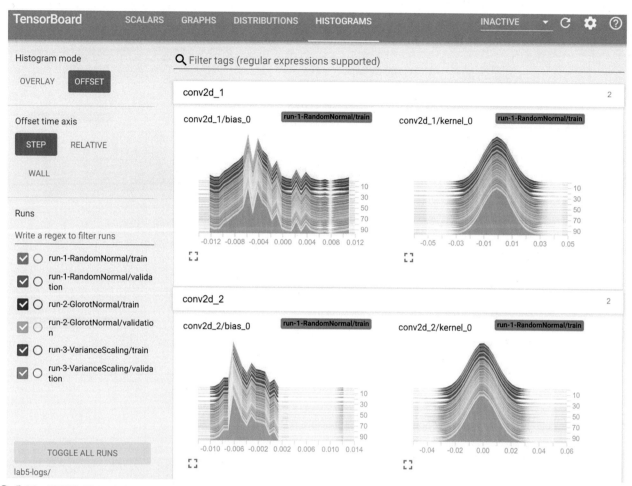

FIG. 5.11 HISTOGRAMS shows the weight changes of Model-1 with normal distribution ($\mu = 0$ and $\sigma = 0.01$). The top left and top right of the figure are the bias distribution and kernel distribution of the second convolution layer, respectively. The bottom left and bottom right of the figure are the bias distribution and kernel distribution of the third convolution layer, respectively.

2. The weights distribution of the Model-2 with Glorot initialization
   (a) Using DISTRIBUTIONS tool

Fig. 5.12 shows the weight changes of the first and second convolutional layers observed by the DISTRIBUTIONS tool when training Model-2 in 100 epochs using the Glorot method for weight initialization. As shown, the longer the training is, the wider the weight distribution becomes, which means that more diverse features can be learned during the training process.

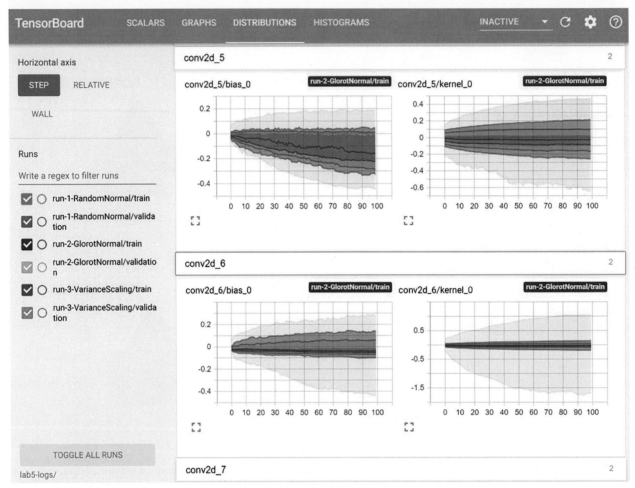

FIG. 5.12   DISTRIBUTIONS shows the weight changes of Model-2 with Glorot initialization. The top left and top right of the figure are the bias distribution and kernel distribution of the first convolution layer, respectively. The bottom left and bottom right of the figure are the bias distribution and kernel distribution of the second convolution layer, respectively.

(b) Using HISTOGRAMS tool

Fig. 5.13 shows the weight changes of the first and second convolutional layers observed by the HISTOGRAMS tool when training Model-2 in 100 epochs using the Glorot method for weight initialization.

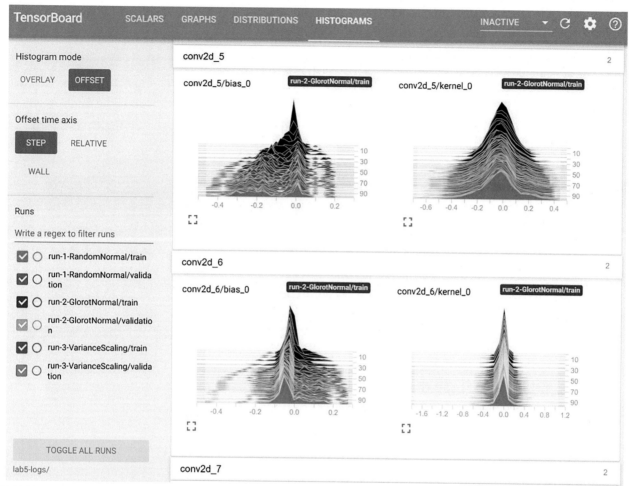

FIG. 5.13 HISTOGRAMS shows the weight changes of Model-2 with Glorot initialization. The top left and top right of the figure are the bias distribution and kernel distribution of the first convolution layer, respectively. The bottom left and bottom right of the figure are the bias distribution and kernel distribution of the second convolution layer, respectively.

3. The weight distribution of the Model-3 with He initialization
   (a) Using DISTRIBUTIONS tool

Fig. 5.14 shows the weight changes of the first and second convolutional layers obtained by using the DISTRIBU-TIONS tool when training Model-3 in 100 epochs using He weight initialization. Compared with the weight distribution of Model-2, the weight distribution Model-3 is wider, which means that Model-3 with He initialization can learn more diverse features than that of Model-2 with Glorot initialization.

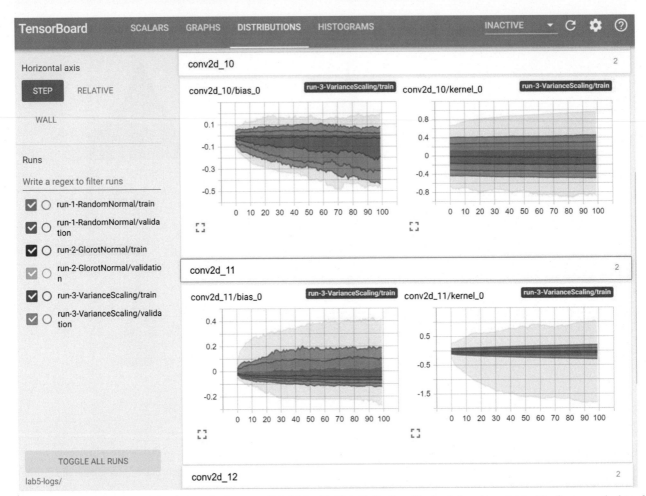

FIG. 5.14 DISTRIBUTIONS shows the weight changes of Model-3 with He initialization. The top left and top right of the figure are the bias distribution and kernel distribution of the first convolution layer, respectively. The bottom left and bottom right of the figure are the bias distribution and kernel distribution of the second convolution layer, respectively.

(b) Using HISTOGRAMS tool

Fig. 5.15 shows the weight changes of the first and second convolutional layers observed by using the HISTO-GRAMS tool when training Model-3 in 100 epochs using He weight initialization.

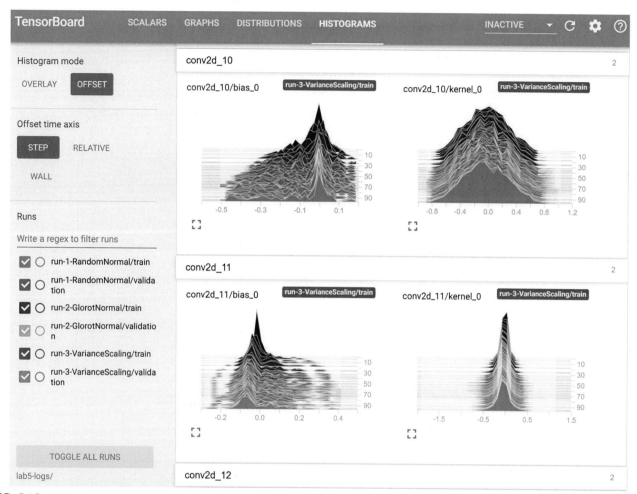

FIG. 5.15 HISTOGRAMS shows the weight changes of Model-3 with He initialization. The top left and top right of the figure are the bias distribution and kernel distribution of the first convolution layer, respectively. The bottom left and bottom right of the figure are the bias distribution and kernel distribution of the second convolution layer, respectively.

5.5 Experiment 2: Verification of batch normalization

In this section, we build a neural network, named Model-4, by adding batch normalization to the convolutional layers of the network architecture in Section 5.4. Table 5.3 shows the architecture of Model-4. This model is trained on the CIFAR-10 dataset to verify the effeteness of batch normalization in improving performance of the neural network.

TABLE 5.3   The architecture of a neural network with batch normalization.

Name	Architecture	Description
Model-4	- Input layer with shape of (32, 32, 3) - Five convolutional layers, followed by batch normalization and ReLu activation function, respectively - One max pooling layer - One flatten layer for flattening the input into a one-dimensional Tensor. - One fully connected layer. - One dropout layer with a discard rate of 50%. - Output fully connected layer with 10 neurons, followed by softmax function.	- Neural network with batch normalization - Batch normalization is placed between the convolutional layer and ReLu activation

■ Building network:

```
inputs = keras.Input(shape=(32, 32, 3))
x = layers.Conv2D(64, (3, 3))(inputs)
x = layers.BatchNormalization()(x)
x = layers.ReLU()(x)
x = layers.MaxPool2D()(x)
x = layers.Conv2D(128, (3, 3))(x)
x = layers.BatchNormalization()(x)
x = layers.ReLU()(x)
x = layers.Conv2D(256, (3, 3))(x)
x = layers.BatchNormalization()(x)
x = layers.ReLU()(x)
x = layers.Conv2D(128, (3, 3))(x)
x = layers.BatchNormalization()(x)
x = layers.ReLU()(x)
x = layers.Conv2D(64, (3, 3))(x)
x = layers.BatchNormalization()(x)
x = layers.ReLU()(x)
x = layers.Flatten()(x)
x = layers.Dense(64)(x)
x = layers.BatchNormalization()(x)
x = layers.ReLU()(x)
x = layers.Dropout(0.5)(x)
outputs = layers.Dense(10, activation='softmax')(x)


model_4 = keras.Model(inputs, outputs, name='model-4')
# Show network architecture
model_4.summary()
```

Result:

```
Model: "model-4"

Layer (type)                    Output Shape              Param #
=================================================================
input_4 (InputLayer)            [(None, 32, 32, 3)]       0

conv2d_15 (Conv2D)              (None, 30, 30, 64)        1792

batch_normalization (BatchNo    (None, 30, 30, 64)        256

re_lu (ReLU)                    (None, 30, 30, 64)        0

max_pooling2d_3 (MaxPooling2    (None, 15, 15, 64)        0

conv2d_16 (Conv2D)              (None, 13, 13, 128)       73856

batch_normalization_1 (Batch    (None, 13, 13, 128)       512

re_lu_1 (ReLU)                  (None, 13, 13, 128)       0

conv2d_17 (Conv2D)              (None, 11, 11, 256)       295168

batch_normalization_2 (Batch    (None, 11, 11, 256)       1024

re_lu_2 (ReLU)                  (None, 11, 11, 256)       0

conv2d_18 (Conv2D)              (None, 9, 9, 128)         295040

batch_normalization_3 (Batch    (None, 9, 9, 128)         512

re_lu_3 (ReLU)                  (None, 9, 9, 128)         0

conv2d_19 (Conv2D)              (None, 7, 7, 64)          73792

batch_normalization_4 (Batch    (None, 7, 7, 64)          256

re_lu_4 (ReLU)                  (None, 7, 7, 64)          0

flatten_3 (Flatten)             (None, 3136)              0

dense_6 (Dense)                 (None, 64)                200768

batch_normalization_5 (Batch    (None, 64)                256

re_lu_5 (ReLU)                  (None, 64)                0

dropout_3 (Dropout)             (None, 64)                0

dense_7 (Dense)                 (None, 10)                650
=================================================================
Total params: 943,882
Trainable params: 942,474
Non-trainable params: 1,408
```

■ Set Callback function:

```
model_dir = 'lab5-logs/models/'  # Create storage directory
# Save training log
log_dir = os.path.join('lab5-logs', 'run-4-batchnormalization')
model_cbk = keras.callbacks.TensorBoard(log_dir=log_dir)
# Store the best model weights
model_mckp = keras.callbacks.ModelCheckpoint(model_dir + '/run-4-best-model.h5',
                    monitor='val_categorical_accuracy',
                    save_best_only=True,
                    mode='max')
```

■ Set the optimizer, loss function, and metric function.

```
model_4.compile(keras.optimizers.Adam(),
        loss=keras.losses.CategoricalCrossentropy(),
        metrics=[keras.metrics.CategoricalAccuracy()])
```

■ Training Model-4

```
history_1 = model_4.fit(train_data,
            epochs=100,
            validation_data=valid_data,
            callbacks=[model_cbk, model_mckp])
```

Reuslt：

```
Epoch 95/100
704/704 [==============================] - 17s 24ms/step - loss: 0.4179 - categorical_accuracy: 0.8642 - val_loss: 0.5299
 - val_categorical_accuracy: 0.8418
Epoch 96/100
704/704 [==============================] - 17s 24ms/step - loss: 0.4190 - categorical_accuracy: 0.8652 - val_loss: 0.8964
 - val_categorical_accuracy: 0.7376
Epoch 97/100
704/704 [==============================] - 17s 24ms/step - loss: 0.4236 - categorical_accuracy: 0.8633 - val_loss: 0.4678
 - val_categorical_accuracy: 0.8470
Epoch 98/100
704/704 [==============================] - 17s 25ms/step - loss: 0.4094 - categorical_accuracy: 0.8669 - val_loss: 0.4725
 - val_categorical_accuracy: 0.8548
Epoch 99/100
704/704 [==============================] - 17s 24ms/step - loss: 0.4174 - categorical_accuracy: 0.8640 - val_loss: 0.4594
 - val_categorical_accuracy: 0.8648
Epoch 100/100
704/704 [==============================] - 17s 24ms/step - loss: 0.4061 - categorical_accuracy: 0.8679 - val_loss: 0.4857
 - val_categorical_accuracy: 0.8552
```

■ Verification on the test set

```
loss, acc = model_4.evaluate(test_data)
print('\nModel-4 Accuracy: {}%'.format(acc))
```

Result：Model-4 Accuracy: 0.8593000173568726%

5.6 Comparison of different neural networks

In this section, we summarize and compare the performance of all the neural networks in Chapter 4 and this chapter. Table 5.4 shows the architecture and the accuracy of the models evaluated on the CIFAR-10 dataset. As shown, the convolutional neural network with batch normalization and Glorot weight initialization (Lab 5 Model-4) is superior to the other models in terms of accuracy.

TABLE 5.4 Performance comparison of seven neural networks.

Network architecture	IA	GU	RN	GN	HN	BN	Accuracy (%)
Fully connected neural network (Lab4 Model-1)	✗	✓	✗	✗	✗	✗	44.53
Convolutional Neural Network (Lab4 Model-2)	✗	✓	✗	✗	✗	✗	72.11
Convolutional Neural Network (Lab4 Model-3)	✓	✓	✗	✗	✗	✗	79.80
Convolutional Neural Network (Lab5 Model-1)	✓	✗	✓	✗	✗	✗	10.00
Convolutional Neural Network (Lab5 Model-2)	✓	✗	✗	✓	✗	✗	79.33
Convolutional Neural Network (Lab5 Model-3)	✓	✗	✗	✗	✓	✗	81.18
Convolutional Neural Network (Lab5 Model-4)	✓	✓	✗	✗	✗	✓	85.93

IA, image augmentation; *RN*, random normal distribution initialization; *GU*, Glorot uniform initialization; *GN*, Glorot normal initialization; *HN*, He normal initialization; *BN*, batch normalization.

The training results of the models in Chapter 4 and this chapter can be observed through TensorBorad, as shown in Fig. 5.16. Note that it is required to enter two log files after –logdir when opening TensorBoard.

```
tensorboard --logdir lab4-logs/,lab5-logs
```

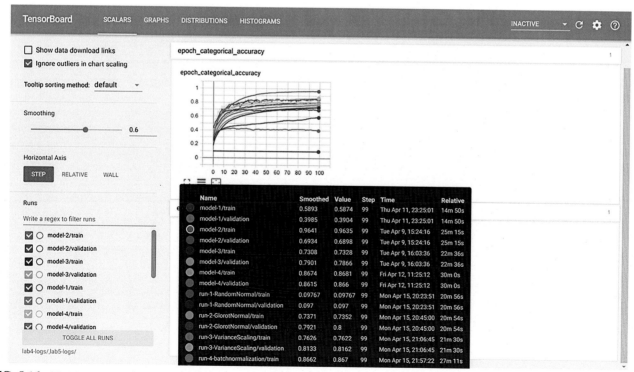

FIG. 5.16 Training result of neural networks on the CIFAR-10 dataset.

References

[1] D.E. Rumelhart, G.E. Hinton, R.J. Williams, Learning representations by back-propagation errors, Nature 323 (1986) 533–536.

[2] D. Mishkin, J. Matas, All you need is a good init, in: Proc. International Conference on Learning Representations, 2016, pp. 3013–3018.

[3] D. Xie, J. Xiong, S. Pu, All you need is beyond a good init: exploring better solution for training extremely deep convolutional neural networks with orthonormality and modulation, Proc. IEEE Conf. Comput. Vis. Pattern Recognit. (2017) 6176–6185.

[4] T. Salimans, D.P. Kingma, Weight normalization: a simple reparameterization to accelerate training of deep neural networks, Adv. Neural Inf. Proces. Syst. (2016) 901–909.

[5] S. Masood, M.N. Doja, P. Chandra, Analysis of weight initialization methods for gradient descent with momentum, in: International Conference on Soft Computing Techniques and Implementations, 2015, pp. 131–136.

[6] X. Glorot, Y. Bengio, Understanding the difficulty of training deep feedforward neural networks, in: Proc. Conf. Artificial Intelligence and Statistics, 2010, pp. 249–256.

[7] K. He, X. Zhang, S. Ren, J. Sun, Delving deep into rectifiers: surpassing human-level performance on ImageNet classification, in: IEEE International Conference on Computer Vision, 2015, pp. 1026–1034.

[8] S. Ioffe, C. Szegedy, Batch normalization: accelerating deep network training by reducing internal covariate shift, in: International Conference on Machine Learning, 2015, pp. 448–456.

[9] G. Huang, Z. Liu, L. Van Der Maaten, K.Q. Weinberger, Densely connected convolutional networks, in: Proceedings of the IEEE Conference on Computer Vision and Pattern Recognition, 2017, pp. 4700–4708.

[10] K. He, X. Zhang, S. Ren, J. Sun, Deep residual learning for image recognition, in: Proceedings of the IEEE Conference on Computer Vision and Pattern Recognition, 2016, pp. 770–778.

[11] T. Le, S. Huang, D. Jaw, Cross-resolution feature fusion for fast hand detection in intelligent homecare systems, IEEE Sens. J. 19 (2019) 4696–4704.

[12] P. Isola, J.-Y. Zhu, T. Zhou, A.A. Efros, Image-to-image translation with conditional adversarial networks, in: Proceedings of the IEEE Conference on Computer Vision and Pattern Recognition, 2017, pp. 1125–1134.

6

Advanced TensorFlow

6.1 Advanced TensorFlow

The previous chapters introduced how to use the tf.keras API to build and train neural network models for linear regression, binary classification, and multi-category classification via the following steps:
- Define the input of the network model: tf.keras.Input()
- Connect the layers of the network: tf.keras.layers()
- Create the network model: tf.keras.Model()
- Set the optimizer, loss function, and metric function: model.compile()
- Train network model: model.fit()

Although the high-level Keras API of TensorFlow makes it is very easy to build and train neural network models, it still has limitations. For example, not all loss functions for all problems are included in the high-level API, so these loss functions are not available to use for training new network models. In this section, we demonstrate advanced Tensor-Flow techniques to show how to customize the API for the following specific purposes.

- Custom network layer: how to create a new network layer
- Custom loss function: how to create a new loss function
- Custom metric function: how to create a new metric function
- Custom callback function: how to create a new callback function

6.1.1 Custom network layer

The types of network layers provided by TensorFlow are quite diverse and include:

- Convolutional layers: Conv1D, Conv2D, Conv3D, SeparableConv2D, DepthwiseConv2D, Conv2DTranspose, and so on
- Pooling layers: MaxPooling1D, MaxPooling2D, AveragePooling2D, GlobalMaxPooling2D, and so on

- Merge layers: Add, Subtract, Multiply, Concatenate, Maximum, and so on
- Activation layers: Rectified linear unit (ReLU), LeakyReLU, PReLU, and so on
- Recurrent layers: RNN, GRU, LSTM, ConvLSTM2D, and so on
- Batch normalization, dropout, and so on

If a type of network layer is not defined in TensorFlow, a new custom one can be created by inheriting the class "tf.keras.layers.Layer." The following is a template for creating a custom layer.

```
class CustomLayer(tf.keras.layers.Layer):
    def __init__(self, **kwargs):
        super(CustomLayer, self).__init__( **kwargs)
        """
        The place to set parameters
        """
    def build(self, input_shape):
        """
        The place to create weights (via add_weight method).
        Parameter:
            input_shape: input size
        """
    def call(self, inputs):
        """
        Define the forward pass (operation) of the network.
        parameter:
            inputs: input data.
        """
    def get_config(self):
        """
        (Optional!) If you want to support serialization, define it here, it will return the layer's construction parameters
        """
```

The official TensorFlow document for Keras layers API is provided at www.tensorflow.org/versions/r2.0/api_docs/python/tf/keras/layers

6.1.2 Custom loss function

There are many loss functions to choose from when designing neural network models for various problems such as classification, recognition, regression, and so on. The loss functions provided by TensorFlow are not sufficient to deal with all problems. Therefore, if a loss function is not available in TensorFlow for addressing a specific problem, it is necessary to define a new one. The following is a template for creating a custom loss function.

```
def custom_loss(y_true, y_pred):
    """

    Define loss function in this place
    Parameter:
        y_true (true value) : expected output
        y_pred (predicted value) : The prediction result of the network。
    """

    return loss
```

The official loss functions provided by TensorFlow can be accessed at www.tensorflow.org/versions/r2.0/api_docs/python/tf/keras/losses

6.1.3 Custom metric function

The metric function is used to evaluate the quality of neural network models, and each metric for each task is different. For example, the intersection-over-union (IoU) metric is employed for semantic segmentation models, the mean average precision (mAP) metric is used for object detection models, and the peak signal-to-noise ratio (PSNR) and structural similarity index (SSIM) metrics are used for image enhancement models such as haze removal model, fog removal model, and so on. If a metric function is not available in TensorFlow, it can be created by inheriting the class "tf.keras.metrics.Metric." The following is a template for creating a custom metric function.

```
class CustomMetrics(tf.keras.metrics.Metric):
    def __init__(self, name='custom_metrics', **kwargs):
        super(CustomMetrics, self).__init__(name=name, **kwargs)
        """

        All state variables used for the metric function need to be defined here.
        Parameter:
            name: The name of the metric function.
        """

    def update_state(self, y_true, y_pred, sample_weight=None):
        """

        Use y_true and y_pred to calculate update state variables.
        Parameter:
            y_true: expected output
            y_pred: predicted value of the network model.
            sample_weight: The weight of the sample, usually used in the sequence model.
        """

    def result(self):
        """

        Use state variables to calculate the final result.
        """

    def reset_states(self):
        """

        Re-initialize the metric function (state variable)
        """
```

The official metric functions provided by TensorFlow can be accessed at www.tensorflow.org/versions/r2.0/api_docs/python/tf/keras/metrics

6.1.4 Custom callback function

A callback is a powerful tool to perform actions at various stages of training neural network models. For example, using callbacks can help to monitor losses and metrics through writing TensorBoard logs after every epoch of training, prevent overfitting by doing early stopping, save the best model weights, and so on. The common callbacks available in TensorFlow are:

- tf.keras.callbacks.ModelCheckpoint: Monitor values and store the best model weights
- tf.keras.callbacks.EarlyStopping: If the monitored values of the model are too long without improvement, the training will be terminated early
- tf.keras.callbacks.ReduceLROnPlateau: If the monitored values of the model are too long without improvement, the learning rate will be reduced
- tf.keras.callbacks.TensorBoard: Record weights, graph, and so on, of the model during the training process

If a callback is not provided by TensorFlow, one can be created by inheriting the class "tf.keras.callbacks.Callback." The following is a template for creating a custom callback.

```
class CustomCallbacks(tf.keras.callbacks.Callback):
    def on_epoch_(begin|end)(self, epoch, logs=None):
        """
        Each epoch is started or stopped, execute this program.
        Parameters:
            epoch: The current epoch.
            logs: Input information such as loss, val_loss, etc in dict format
        """

    def on_(train|test|predict)_begin(self, logs=None):
        """
```

```python
class CustomCallbacks(tf.keras.callbacks.Callback):
    def on_epoch_(begin|end)(self, epoch, logs=None):
        """

        Each epoch is started or stopped, execute this program.
        Parameters:
            epoch: The current epoch.

            logs: Input information such as loss, val_loss, etc in dict format
        """

    def on_(train|test|predict)_begin(self, logs=None):
        """

        fitting, evaluation or prediction tasks starts, execute this program.
        parameters:

            log: Input the information such as loss, val_loss, etc in dict format
        """

    def on_(train|test|predict)_end(self, logs=None):
        """

        At the end of the training, evaluation or prediction stage, execute this program.
        Parameters:

            logs: Input the information such as loss, val_loss, etc in dict format
        """

    def on_(train|test|predict)_batch_begin(self, batch, logs=None):
        """

        This program is executed before the start of each batch of the training, evaluation,
        or prediction tasks.
        Parameters:

            batch: the current batch

            logs: Input the information such as loss, val_loss, etc. in dict format,
        """

    def on_(train|test|predict)_batch_end(self, batch, logs=None):
        """

        This program is executed after each batch of the training, evaluation,
        or prediction is completed.
        Parameters:

            batch: Current batch

            logs: Input the information such as loss, val_loss, etc in dict format
        """
```

The official callback functions provided by TensorFlow can be accessed at www.tensorflow.org/versions/r2.0/ api_docs/python/tf/keras/callbacks

6.2 Using high-level keras API and custom API of TensorFlow

This section presents more detail about the custom API and high-level Keras API of TensorFlow through examples. Each example is implemented using both methods to show the benefits of each one. While the high-level Keras API was optimized and easy to use, the custom API is very flexible in building neural networks.

6.2.1 Network layer

In this section, we create a convolutional layer using the high-level Keras API and custom network layer of Tensor-Flow. The parameters of the convolution layer are:

- Number of kernels: 64
- Kernel size: 3×3
- Strides: 1
- Padding: valid
- Activation function: ReLU
- Weight initialization: Glorot initialization
- Bias initialization: zeros

The convolution layer created by the following two methods have the same function.

a) First method: Using the high-level Keras API of TensorFlow.

```python
tf.keras.layers.Conv2D(64, 3, activation='relu', kernel_initializer='glorot_uniform')
```

b) Second method: Using a custom network layer of TensorFlow.

```python
class CustomConv2D(tf.keras.layers.Layer):
    def __init__(self, filters, kernel_size, strides=(1, 1), padding="VALID", **kwargs):
        super(CustomConv2D, self).__init__(**kwargs)
        self.filters = filters
        self.kernel_size = kernel_size
        self.strides = (1, *strides, 1)
        self.padding = padding

    def build(self, input_shape):
        kernel_h, kernel_w = self.kernel_size
        input_dim = input_shape[-1]
        # create the weights
        self.w = self.add_weight(name='kernel',
                    shape=(kernel_h, kernel_w, input_dim, self.filters),
                    initializer='glorot_uniform', # weight initialization
                    trainable=True) # Set whether weight can be trained
        # Create the biases
        self.b = self.add_weight(name='bias',
                    shape=(self.filters,),
                    initializer='zeros', # bias initialization
                    trainable=True) # Set whether weight can be trained

    def call(self, inputs):
        x = tf.nn.conv2d(inputs, self.w, self.strides, padding=self.padding) # Convolution operation
        x = tf.nn.bias_add(x, self.b)
        x = tf.nn.relu(x) # using Relu activation function
        return x
```

6.2.2 Loss function

This section establishes the categorical cross-entropy (CCE) loss function using both high-level Keras API and custom loss function of TensorFlow. The formula for CCE is:

$$CCE = -\frac{\sum_{i=1}^{N}\sum_{j=1}^{C} y_{i,j}\log\left(f(\hat{y}_{i,j})\right)}{N}$$

where y is expected output, $\hat{y}$ is prediction result of the network model, f is softmax function, C is number of categories, and N is the amount of data in a batch.

The loss functions created by the following two methods have the same function.

a) First method: Using the high-level Keras API of TensorFlow.

```
tf.keras.losses.CategoricalCrossentropy()
```

b) Second method: Using a custom loss function of TensorFlow.

```
def custom_categorical_crossentropy(y_true, y_pred):
    x = tf.reduce_mean(-tf.reduce_sum(y_true * tf.log(y_pred), reduction_indices=[1]))
    return x
```

Although the custom categorical cross-entropy (CCE) loss function created by the second method can work, it will be replaced with the official function "tf.nn.softmax_cross_entropy" in practice. The reason is that the official CCE API has been optimized, and thus it is faster and more stable in training.

```
def custom_categorical_crossentropy(y_true, y_pred):
    x = tf.nn.softmax_cross_entropy_with_logits(labels=y_true, logits=y_pred)
    return x
```

6.2.3 Metric function

A metric is a function that is employed to judge the performance of the model. TensorFlow provides a lot of metrics for performance judging such as binary accuracy, categorical accuracy, sparse categorical accuracy, top-k categorical accuracy, sparse_top_k categorical accuracy, and so on. In this section, we create the categorical accuracy which calculates how often predictions matches one-hot labels by using the high-level Keras API and custom metric function of TensorFlow.

The metric functions created by the following two methods have the same function.

a) First method: Using the high-level Keras API of TensorFlow.

```
tf.keras.metrics.CategoricalAccuracy()
```

b) Second method: Using custom metric functions of TensorFlow.

```python
class CustomCategoricalAccuracy(tf.keras.metrics.Metric):
    def __init__(self, name='custom_catrgorical_accuracy', **kwargs):
        super(CustomCategoricalAccuracy, self).__init__(name=name, **kwargs)
        # the number of correct predictions
        self.correct = self.add_weight('correct_numbers', initializer='zeros')
        # the amount of all data
        self.total = self.add_weight('total_numbers', initializer='zeros')

        def update_state(self, y_true, y_pred, sample_weight=None):
            (y_true is represented by using One-hot encoding)
            # get the index with the largest value
            y_true = tf.argmax(y_true, axis=-1)
            # get the index with the largest value
            y_pred = tf.argmax(y_pred, axis=-1)
            # Compare whether the prediction result is correct, true will return True (correct),
            false will return False (error)
            values = tf.equal(y_true, y_pred)
```

```python
            ## Convert to floating point: True (correct) = 1.0, False (false) = 0.0
            values = tf.cast(values, tf.float32)
            # Computes the sum of elements
            values_sum = tf.reduce_sum(values)
            num_values = tf.cast(tf.size(values), tf.float32)
            self.correct.assign_add(values_sum)  # Update the total number of correct
            predictions
            self.total.assign_add(num_values)  # Total amount of updated data

    def result(self):
        # Calculate accuracy
        return tf.math.divide_no_nan(self.correct, self.total)

    def reset_states(self):
        # Variables will be reinitialized after each Epoch
        self.correct.assign(0.)
        self.total.assign(0.)
```

Supplementary explanation

An example of categorical accuracy calculation.

```
# y_true (answer, One-hot encoding) is a two-dimensional Tensor, Tensor shape=(3, 10)
y_true = tf.constant([[0, 0, 0, 1, 0, 0, 0, 0, 0, 0],
                      [0, 0, 0, 0, 1, 0, 0, 0, 0, 0],
                      [1, 0, 0, 0, 0, 0, 0, 0, 0, 0]])
# y_pred (prediction output, probability score) is a two-dimensional Tensor, shape=(3, 10)
y_pred = tf.constant([[.1, 0, 0, .7, 0, 0, 0, 0.2, 0, 0],
                      [0, 0, 0, 0, .8, .1, 0, 0, .1, 0],
                      [.3, 0, 0, 0, 0, 0, 0, .1, .6, 0]])
#   Get the index with the largest value ([3, 4, 0])
y_true = tf.argmax(y_true, axis=-1)
# Get the index with the largest value ([3, 4, 8])
y_pred = tf.argmax(y_pred, axis=-1)
# check whether the prediction result is correct
# tf.equal([3, 4, 0], [3, 4, 8]) →[True, True, False]
values = tf.equal(y_true, y_pred)
# Convert to floating point (True=1.0, False=0.0)
# tf.cast([True, True, False], tf.float32) →[1., 1., 0.]
values = tf.cast(values, tf.float32)
# tf.reduce_sum([1., 1., 0.]) →2.0
values_sum = tf.reduce_sum(values)
# get the number of data in this batch, # tf.size([1. 1. 0.], out_type=tf.float32) →3.0
num_values = tf.cast(tf.size(values), tf.float32)
```

6.2.4 Callback function

This section introduces two methods to save the best model weights of the neural network during the training process. The first method employs a callback of the high-level Keras API, while the second method adopts a custom callback function of TensorFlow.

The callbacks created by the following two methods have the same function.

    a) First method: Using the high-level Keras API of TensorFlow.

```python
tf.keras.callbacks.ModelCheckpoint('logs/models/save.h5')
```

    b) Second method: Use a custom callback function of TensorFlow.

```python
class SaveModel(tf.keras.callbacks.Callback):
    def __init__(self, weights_file, monitor='loss', mode='min', save_weights_only=False):
        super(SaveModel, self).__init__()
        # Set storage path
        self.weights_file = weights_file
        # Set the value to be monitored
        self.monitor = monitor
        # Set the monitoring value to "bigger is better" or "smaller is better"
        # Ex: The monitoring value is loss, it must be set to 'min', if the monitoring value
        # is Accuracy, then set to 'max'.
        self.mode = mode
        # Save model weights only or Save entire network model (including Layer,
        # Compile, etc.)
        self.save_weights_only = save_weights_only
        if mode == 'min':
            # Set best to infinity
            self.best = np.Inf
        else:
            # Set best to negative infinity
            self.best = -np.Inf

    # Function for saving network model
    def save_model(self):
        if self.save_weights_only:
            # Only save model weights
            self.model.save_weights(self.weights_file)
        else:
            # Save the entire network model (including Layer, Compile, etc.)
            self.model.save(self.weights_file)

    def on_epoch_end(self, epoch, logs=None):
```

```python
        # Read monitored values from logs
        monitor_value = logs.get(self.monitor)
        # If the monitored values decrease or increase (depending on the mode setting)
        # , save the network model
        if self.mode == 'min' and monitor_value < self.best:
            self.save_model()
            self.best = monitor_value
        elif self.mode == 'max' and monitor_value > self.best:
            self.save_model()
            self.best = monitor_value
```

Fig. 6.1 shows an example of a saved model weights file produced by Keras API of TensorFlow.

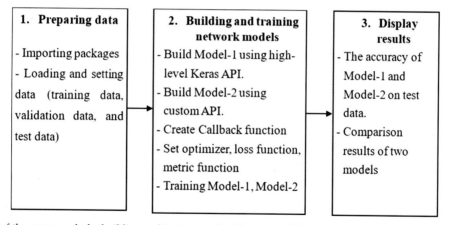

FIG. 6.1 Saved model weights file.

6.3 Experiment: implementation of two network models using high-level keras API and custom API

In this section, we present two scenarios of building neural networks for multi-category classification. In the first scenario, we employ the high-level Keras API of TensorFlow to create the network model, called Model-1. In the second scenario, we use the custom API of TensorFlow to build the network model, called Model-2. Model-1 and Model-2 are trained separately on the CIFAR-10 dataset for comparing performance.

Fig. 6.2 is a flowchart of the source code for building Model-1 and Model-2.

1. Preparing data	2. Building and training network models	3. Display results
- Importing packages - Loading and setting data (training data, validation data, and test data)	- Build Model-1 using high-level Keras API. - Build Model-2 using custom API. - Create Callback function - Set optimizer, loss function, metric function - Training Model-1, Model-2	- The accuracy of Model-1 and Model-2 on test data. - Comparison results of two models

FIG. 6.2 Flowchart of the source code for building multi-category classification models using high-level Keras API and custom API.

1. Preparing data

a) Import packages

```python
import os
import numpy as np
import pandas as pd
import tensorflow as tf
import tensorflow_datasets as tfds
from tensorflow import keras
from tensorflow.keras import layers
from tensorflow.keras import initializers
# Import parse_aug_fn and parse_fn functions from the preprocessing.py file
from preprocessing import parse_aug_fn, parse_fn
```

b) Loading and setting data

- Loading data:

```python
# Divide training data 9: 1 (9 for training and 1 for validation) train_split,
valid_split = ['train[:90%]', 'train[90%:]']
# get the training data
train_data, info = tfds.load("cifar10", split=train_split, with_info=True)
# get the valid data
valid_data = tfds.load("cifar10", split=valid_split)
#get the test data
test_data = tfds.load("cifar10", split=tfds.Split.TEST)
# Get the category of CIFAR-10 dataset
class_name = info.features['label'].names
```

- Setting data:

```python
AUTOTUNE = tf.data.experimental.AUTOTUNE  # Automatic adjustment mode
batch_size = 64  # Batch size
train_num = int(info.splits['train'].num_examples / 10) * 9  # Number of training data


# Shuffle the training data
train_data = train_data.shuffle(train_num)
# Training data
train_data = train_data.map(map_func=parse_aug_fn, num_parallel_calls=AUTOTUNE)
# Set batch size and turn on prefetch mode
train_data = train_data.batch(batch_size).prefetch(buffer_size=AUTOTUNE)


# Validation data
valid_data = valid_data.map(map_func=parse_fn, num_parallel_calls=AUTOTUNE)
# Set batch size and turn on prefetch mode
valid_data = valid_data.batch(batch_size).prefetch(buffer_size=AUTOTUNE)


# Test data
test_data = test_data.map(map_func=parse_fn, num_parallel_calls=AUTOTUNE)
#Set batch size and turn on prefetch mode
test_data = test_data.batch(batch_size).prefetch(buffer_size=AUTOTUNE)
```

2. Building and training network model

Model-1 and Model-2 have the same architecture, as shown in Table 6.1. While Model-1 is constructed using the high-level Keras API of TensorFlow, Model-2 is built using the custom API of TensorFlow.

TABLE 6.1 The architecture of Model-1 and Model-2.

Name	Architecture	Description
Model-1	- Input layer with shape of (32,32,3) - Five convolutional layers, followed by rectified linear unit (ReLU) activation functions - One max pooling layer - One flatten layer for flattening the input into a one-dimensional Tensor - One fully connected layer - One dropout layer with a discard rate of 50% - Output fully connected layer with 10 neurons	Using the high-level Keras API to build and train the network model
Model-2	- Input layer with shape of (32, 32, 3) - Five convolutional layers, followed by ReLU activation functions - One max pooling layer - One flatten layer for flattening the input into a one-dimensional Tensor - One fully connected layer - One dropout layer with a discard rate of 50% - Output fully connected layer with 10 neurons	Using the custom API of TensorFlow to build and train the network model

a) Model-1: Using high-level Keras API of TensorFlow.

■ Build network model:

```
inputs = keras.Input(shape=(32, 32, 3))
x = layers.Conv2D(64, 3, activation='relu', kernel_initializer='glorot_uniform')(inputs)
x = layers.MaxPool2D()(x)
```

```
x = layers.Conv2D(128, 3, activation='relu', kernel_initializer='glorot_uniform')(x)
x = layers.Conv2D(256, 3, activation='relu', kernel_initializer='glorot_uniform')(x)
x = layers.Conv2D(128, 3, activation='relu', kernel_initializer='glorot_uniform')(x)
x = layers.Conv2D(64, 3, activation='relu', kernel_initializer='glorot_uniform')(x)
x = layers.Flatten()(x)
x = layers.Dense(64, activation='relu')(x)
x = layers.Dropout(0.5)(x)
outputs = layers.Dense(10)(x)
# Create a network model (connect all the network layers that pass through from input to
output)
model_1 = keras.Model(inputs, outputs, name='model-1')
model_1.summary()
```

Result:

```
Model: "model 1"

Layer (type)                   Output Shape            Param #
=================================================================
input_5 (InputLayer)           [(None, 32, 32, 3)]     0

conv2d_7 (Conv2D)              (None, 30, 30, 64)      1792

max_pooling2d_4 (MaxPooling2   (None, 15, 15, 64)      0

conv2d_8 (Conv2D)              (None, 13, 13, 128)     73856

conv2d_9 (Conv2D)              (None, 11, 11, 256)     295168

conv2d_10 (Conv2D)             (None, 9, 9, 128)       295040

conv2d_11 (Conv2D)             (None, 7, 7, 64)        73792

flatten_4 (Flatten)            (None, 3136)            0

dense_8 (Dense)                (None, 64)              200768

dropout_4 (Dropout)            (None, 64)              0

dense_9 (Dense)                (None, 10)              650
=================================================================
Total params: 941,066
Trainable params: 941,066
Non-trainable params: 0
```

■ Create callback function:

```
# Save training log
logs_dirs = 'lab6-logs'
model_cbk = keras.callbacks.TensorBoard(log_dir='lab6-logs')
# create a path to save models
model_dirs = logs_dirs + '/models'
os.makedirs(model_dirs, exist_ok=True)
```

```
save_model = keras.callbacks.ModelCheckpoint(model_dirs + '/save.h5',
                                   monitor='val_catrgorical_accuracy',
                                   mode='max')
```

- Set the optimizer, loss function, and metric function:

```
model_1.compile(keras.optimizers.Adam(),
          # Since Softmax is not used in the output of the model, set from_logits to True
          loss=keras.losses.CategoricalCrossentropy(from_logits=True),
          metrics=[keras.metrics.CategoricalAccuracy()])
```

- Training Model-1:

```
model_1.fit(train_data,
      epochs=100,
      validation_data=valid_data,
      callbacks=[model_cbk, save_model])
```

Result:

```
Epoch 96/100
704/704 [==============================] - 12s 17ms/step - loss: 0.8100 - categorical_accuracy: 0.7326 - val_loss: 0.6121
 - val_categorical_accuracy: 0.7950
Epoch 97/100
704/704 [==============================] - 12s 17ms/step - loss: 0.8067 - categorical_accuracy: 0.7320 - val_loss: 0.6470
 - val_categorical_accuracy: 0.7874
Epoch 98/100
704/704 [==============================] - 12s 17ms/step - loss: 0.7964 - categorical_accuracy: 0.7348 - val_loss: 0.6376
 - val_categorical_accuracy: 0.7856
Epoch 99/100
704/704 [==============================] - 12s 17ms/step - loss: 0.7924 - categorical_accuracy: 0.7360 - val_loss: 0.6556
 - val_categorical_accuracy: 0.7892
Epoch 100/100
704/704 [==============================] - 12s 17ms/step - loss: 0.7985 - categorical_accuracy: 0.7351 - val_loss: 0.6480
 - val_categorical_accuracy: 0.7882

<tensorflow.python.keras.callbacks.History at 0x7f8288443eb8>
```

b) Model-2: Using custom API of TensorFlow

- Create CustomConv2D convolution layer:

```
class CustomConv2D(tf.keras.layers.Layer):
    def __init__(self, filters, kernel_size, strides=(1, 1), padding="VALID", **kwargs):
        super(CustomConv2D, self).__init__(**kwargs)
        self.filters = filters
        self.kernel_size = kernel_size
        self.strides = (1, *strides, 1)
        self.padding = padding

    def build(self, input_shape):
        kernel_h, kernel_w = self.kernel_size
        input_dim = input_shape[-1]
```

```python
# create the weight
self.w = self.add_weight(name='kernel',
            shape=(kernel_h, kernel_w, input_dim, self.filters),
            initializer='glorot_uniform',  # Set weight initialization
            trainable=True)  # Set whether this weight can be trained
# Create the biases
self.b = self.add_weight(name='bias',
            shape=(self.filters,),
            initializer='zeros',  # bias initialization
            trainable=True)  # Set whether this weight can be trained


def call(self, inputs):
    x = tf.nn.conv2d(inputs, self.w, self.strides, padding=self.padding)  # Convolution
operation
    x = tf.nn.bias_add(x, self.b)
    x = tf.nn.relu(x)  # using Relu activation function
    return x
```

- Build network model:

```python
inputs = keras.Input(shape=(32, 32, 3))
x = CustomConv2D(64, (3, 3))(inputs)
x = layers.MaxPool2D()(x)
x = CustomConv2D(128, (3, 3))(x)
x = CustomConv2D(256, (3, 3))(x)
x = CustomConv2D(128, (3, 3))(x)
x = CustomConv2D(64, (3, 3))(x)
x = layers.Flatten()(x)
x = layers.Dense(64, activation='relu')(x)
x = layers.Dropout(0.5)(x)
outputs = layers.Dense(10)(x)
# Create a network model
model_2 = keras.Model(inputs, outputs, name='model-2')
model_2.summary()
```

Result:

```
Model: "model 2"

Layer (type)                    Output Shape              Param #
=================================================================
input_6 (InputLayer)            [(None, 32, 32, 3)]       0

custom_conv2d_15 (CustomConv    (None, 30, 30, 64)        1792

max_pooling2d_5 (MaxPooling2    (None, 15, 15, 64)        0

custom_conv2d_16 (CustomConv    (None, 13, 13, 128)       73856

custom_conv2d_17 (CustomConv    (None, 11, 11, 256)       295168

custom_conv2d_18 (CustomConv    (None, 9, 9, 128)         295040

custom_conv2d_19 (CustomConv    (None, 7, 7, 64)          73792

flatten_5 (Flatten)             (None, 3136)              0

dense_10 (Dense)                (None, 64)                200768

dropout_5 (Dropout)             (None, 64)                0

dense_11 (Dense)                (None, 10)                650
=================================================================
Total params: 941,066
Trainable params: 941,066
Non-trainable params: 0
```

Supplementary explanation

Model-1 and Model-2 do not use the softmax activation function at the output layer because the built-in loss function (keras.losses.CategoricalCrossentropy()) of Model-1 has an argument "from_logits" was set to "True" and the custom loss function of Model-2 utilized the "tf.nn.softmax_cross_entropy_with_logits()" function, which already adopted softmax for calculating CCE loss. Therefore, if the softmax activation function is attached at the output layer of the models, the output value will be calculated two times through the softmax function.

- Create SaveModel callback function:

```
class SaveModel(tf.keras.callbacks.Callback):
    def __init__(self, weights_file, monitor='loss', mode='min',save_weights_only=False):
        super(SaveModel, self).__init__()
        # set storage path
        self.weights_file = weights_file
        # Set the value to be monitored
        self.monitor = monitor
        # Set the monitoring value: "bigger is better" or "smaller is better"
        self.mode = mode
        # Save model weights only or Save entire network model (including Layer,
```

```
    Compile, etc.)
    self.save_weights_only = save_weights_only
    if mode == 'min':
        # Set best to infinity
        self.best = np.Inf
    else:
        # Set best to negative infinity
        self.best = -np.Inf


# Function for saving network model
def save_model(self):
    if self.save_weights_only:
        self.model.save_weights(self.weights_file)
    else:
        self.model.save(self.weights_file)


def on_epoch_end(self, epoch, logs=None):
    # Read monitored values from logs
    monitor_value = logs.get(self.monitor)
    # If the monitored values decrease or increase (depending on your mode setting),
    save the network model
    if self.mode == 'min' and monitor_value < self.best:
        self.save_model()
        self.best = monitor_value
    elif self.mode == 'max' and monitor_value > self.best:
        self.save_model()
        self.best = monitor_value
```

- Create Callback function: Using custom callback function to save model weights:

```
# Save training log
logs_dirs = 'lab6-logs'
model_cbk = keras.callbacks.TensorBoard(log_dir='lab6-logs')

# create a storage path
model_dirs = logs_dirs + '/models'
os.makedirs(model_dirs, exist_ok=True)
# Custom callback function
custom_save_model = SaveModel(model_dirs + '/custom_save.h5',
```

```
                    monitor='val_custom_catrgorical_accuracy',
                    mode='max',
                    save_weights_only=True)
```

- Create a custom_categorical_crossentropy loss function:

```
def custom_categorical_crossentropy(y_true, y_pred):
    x = tf.nn.softmax_cross_entropy_with_logits(labels=y_true, logits=y_pred)
    return x
```

■ Create a CustomCategoricalAccuracy metric function:

```python
class CustomCategoricalAccuracy(tf.keras.metrics.Metric):
    def __init__(self, name='custom_catrgorical_accuracy', **kwargs):
        super(CustomCategoricalAccuracy, self).__init__(name=name, **kwargs)
        # Record the number of correct predictions
        self.correct = self.add_weight('correct_numbers', initializer='zeros')
        # Record the amount of all data
        self.total = self.add_weight('total_numbers', initializer='zeros')

    def update_state(self, y_true, y_pred, sample_weight=None):
        #y_true is represented by One-hot encoding
        #Get the index with the largest value
        # EX:tf.argmax([[0,0,0,1,0,0,0,0,0]],axis=-1)=[4]
        y_true = tf.argmax(y_true, axis=-1)
        # Get the index with the largest value
        y_pred = tf.argmax(y_pred, axis=-1)
        # check whether the predicted result is correct, true will return True (correct),
        # false will return False (error)
        values = tf.equal(y_true, y_pred)
        # Convert to floating point: True (correct) = 1.0, False (false) = 0.0
        values = tf.cast(values, tf.float32)
        # tf.reduce_sum([1. , 1. , 0.])→2.0
        values_sum = tf.reduce_sum(values)
        # tf.size([1. 1. 0.], out_type=tf.float32)→3.0
        num_values = tf.cast(tf.size(values), tf.float32)
        self.correct.assign_add(values_sum)
        self.total.assign_add(num_values)

    def result(self):
```

```
    # Calculate the accuracy
    return tf.math.divide_no_nan(self.correct, self.total)


def reset_states(self):
    # Variables will be reinitialized after each Epoch
    self.correct.assign(0.)
    self.total.assign(0.)
```

- Set the optimizer, loss function, and metric function:

```
model_2.compile(keras.optimizers.Adam(),
        loss=custom_categorical_crossentropy,    # Custom loss function
        metrics=[CustomCategoricalAccuracy()])  # Custom metric funtion
```

- Training Model-2:

```
model_2.fit(train_data,
        epochs=100,
        validation_data=valid_data,
        callbacks=[model_cbk, custom_save_model])
```

Result:

```
Epoch 96/100
704/704 [==============================] - 12s 17ms/step - loss: 0.7810 - custom_catrgorical_accuracy: 0.7374 - val_loss:
0.6471 - val_custom_catrgorical_accuracy: 0.7922
Epoch 97/100
704/704 [==============================] - 12s 17ms/step - loss: 0.7745 - custom_catrgorical_accuracy: 0.7422 - val_loss:
0.6365 - val_custom_catrgorical_accuracy: 0.7966
Epoch 98/100
704/704 [==============================] - 12s 17ms/step - loss: 0.7777 - custom_catrgorical_accuracy: 0.7420 - val_loss:
0.6332 - val_custom_catrgorical_accuracy: 0.7954
Epoch 99/100
704/704 [==============================] - 12s 17ms/step - loss: 0.7803 - custom_catrgorical_accuracy: 0.7416 - val_loss:
0.6337 - val_custom_catrgorical_accuracy: 0.8000
Epoch 100/100
704/704 [==============================] - 12s 17ms/step - loss: 0.7669 - custom_catrgorical_accuracy: 0.7473 - val_loss:
0.6622 - val_custom_catrgorical_accuracy: 0.7946

<tensorflow.python.keras.callbacks.History at 0x7f834551c208>
```

3. Displaying results.

- Load model weights of two Model-1 and Model-2:

```
model_1.load_weights(model_dirs+'/save.h5')
model_2.load_weights(model_dirs+'/custom_save.h5')
```

- Verification on test data:

```
loss_1, acc_1 = model_1.evaluate(test_data)
loss_2, acc_2 = model_2.evaluate(test_data)
loss = [loss_1, loss_2]
acc = [acc_1, acc_2]
dict = {"Loss": loss, "Accuracy": acc}
pd.DataFrame(dict)
```

Result:

	Loss	Accuracy
0	0.661652	0.7897
1	0.627981	0.8018

The experimental results show that the accuracy of both models is close. This confirms that using custom API of TensorFlow for designing neural networks can achieve a similar result as using the high-level Keras API while at the same time being more flexible.

7

Advanced TensorBoard

7.1 Advanced TensorBoard

TensorBoard is an official toolkit of TensorFlow that provides the visualization needed for designing neural networks and machine learning experimentation. Chapters 2 and 5 introduced the use of four visualization functions including Scalars, Graphs, Distributions, and Histograms, which only need to use tf.keras.callbacks.TensorBoard for completing settings. However, the tf.keras.callbacks.TensorBoard function in the high-level Keras API is limited in providing information in various states and statistics of the neural networks during the training process. For example, only key metrics such as loss and accuracy, and how they change as training process can be observed in the Scalars dashboard of TensorBoard. If we want to observe and analyze the output result of the network models during training such as image, text, and so on, it is required to use a custom function. This chapter introduces how to write summary data and display on TensorBoard with the "tf.summary" module of TensorFlow, which provides the following APIs:

- tf.summary.scalar: write a scalar summary such as loss value, accuracy value, or learning rate
- tf.summary.image: write an image summary
- tf.summary.text: write a text summary
- tf.summary.audio: write an audio summary
- tf.summary.histogram: write a histogram summary

The following section presents the use of "tf.summary" module.

- Creates a summary file writer for the given log directory:

```
summary_writer = tf.summary.create_file_writer('lab7-logs-summary')
```

- Write the log file:

```
with summary_writer.as_default():
    tf.summary.(scalar | image | text | audio | histogram)(...)   # type of data need to write
```

- View results on TensorBoard: Open TensorBoard through the command line to view the results, as shown in Fig. 7.1. Because data have not been written to the log file, no dashboards are active for the data.

```
tensorboard --logdir lab7-logs-summary
```

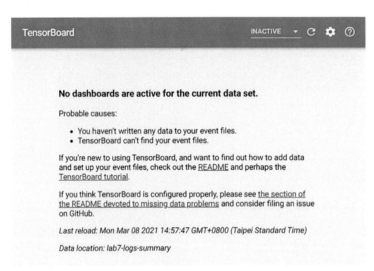

FIG. 7.1   TensorBoard interface.

Details of the "tf.summary" module can be accessed at: www.tensorflow.org/versions/r2.0/api_docs/python/tf/summary

7.1.1 tf.summary.scalar

The "tf.summary.scalar" is an API of the "tf.summary" module used to write a scalar summary such as loss values, accuracy values, and so on. These values can be displayed through the scalars dashboard of TensorBoard for observing and monitoring. Fig. 7.2 presents the loss and mean absolute error value changes of a neural network during training obtained by using the function "tf.keras.callbacks.TensorBoard" of the high-level Keras API. If other scalar values need to be written and monitored, it is suggested to use "tf.summary.scalar". The syntax of "tf.summary.scalar" is expressed as follows:

tf.summary.scalar (name, data, step=None, description=None)

Arguments:

- name: a name for the summary.
- data: a real numeric scalar value, it can be converted to a float32 tensor.
- step: explicit monotonic step value.
- description: optional description for the summary.

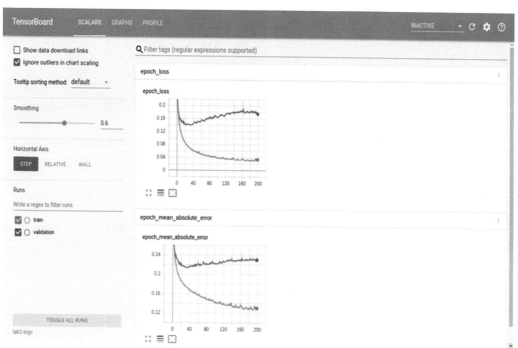

FIG. 7.2 The scalars dashboard.

The following is an example of using the "tf.summary.scalar" API to write and visualize the Sine function on TensorBoard. Fig. 7.3 shows the results.

```
# Generate 100 points linearly between 0 and 2π
x = np.linspace(0, 2 * np.pi , 100)
# using Sin function to generate data
data = np.sin(x)
with summary_writer.as_default():
    for i, y in enumerate(data):
        tf.summary.scalar('sin', y, step=i)
```

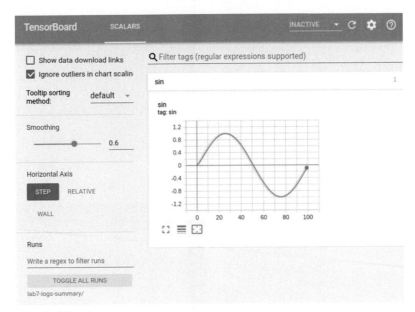

FIG. 7.3   The Sine on the scalars dashboard.

7.1.2 tf.summary.image

"tf.summary.image" is an API of the "tf.summary" module used for writing an image summary. Its result can display on the images dashboard of TensorBoard. The syntax of "tf.summary.image" is expressed as follows:

> tf.summary.image(name, data, step=None, max_outputs=3, description=None)

Arguments:

- name: a name for the summary.
- data: a tensor with shape (n,h,w,c), where n,h,w, and c are the number of images, height, width, and the number of channels of images, respectively.
- step: explicit monotonic step value.
- max_outputs: a positive integer number, indicating the number of images is emitted at each step, the default value is set to 3.
- description: optional description for the summary.

The following program examples explain how to use the "tf.summary.image" API. Please download the test image from the link below and unzip it under the root directory. https://drive.google.com/open?id=1cC45twI3a5AkBYYE6Qb3ZV3uCPlw9eL6

- First read an image (airplane.png)

```
# Create a function to read image
def read_img(file):
    image_string = tf.io.read_file(file)  # read file
    # decode the imported file
    image_decode = tf.image.decode_image(image_string)
    return image_decode


img = read_img('image/airplane.png')  #read image
plt.imshow(img)  # display the image
```

Result:

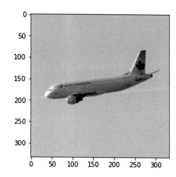

- Write an image to a log file and display it on TensorBoard, as shown in Fig. 7.4

```
with summary_writer.as_default():
    tf.summary.image("Airplane", [img], step=0)
```

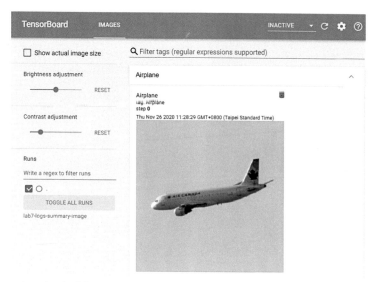

FIG. 7.4 Observing an image written by the "tf.summary.image" on the images dashboard.

- Five images including airplane_zoom.png, plane_flip.png, plane_color.png, plane_rot.png, and plane.png are written to the log file and displayed on TensorBoard at once, as shown in Fig. 7.5

```
img_files = [airplane_zoom.png, airplane_flip.png, airplane_color.png, airplane_rot.png,
        airplane.png]  # create an image array.
imgs = []
for file in imgs:
    imgs.append(read_img('image/'+file))  # read image and store it in the array

with summary_writer.as_default():
    #emit five images
    tf.summary.image("Airplane Augmentation", imgs, max_outputs=5, step=0)
```

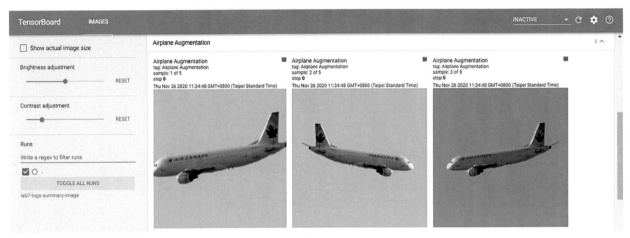

FIG. 7.5 Observing multiple images written by "tf.summary.image" on the images dashboard.

- Write five images including airplane_zoom.png, airplane_flip.png, airplane_color.png, airplane_rot.png, and airplane.png to the log file in separate steps and display on TensorBoard, as shown in Fig. 7.6

```
with summary_writer.as_default():
    # writing in separate step
    for i, img in enumerate(imgs):
        tf.summary.image("Save image each step", [img], step=i)
```

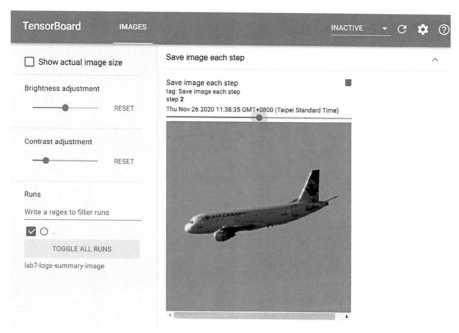

FIG. 7.6 Observing image written by "tf.summary.image" in separate steps on the images dashboard.

7.1.3 tf.summary.text

"tf.summary.text" is an API of the "tf.summary" module responsible for writing text summary. Its results can display through the text dashboard of TensorBoard. The syntax of "tf.summary.text" is expressed as follows:

tf.summary.text(name, data, step=None, description=None)

Arguments:

- name: a name for the text summary
- data: a string tensor value
- step: explicit monotonic step value
- description: optional description for the text summary

The following program example explains how to use "tf.summary.text" API.

- Write text data to a log file and display it on TensorBoard, as shown in Fig. 7.7

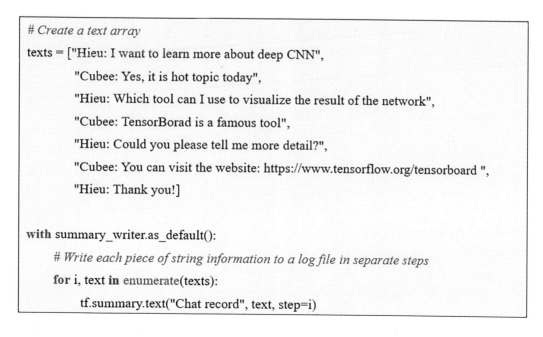

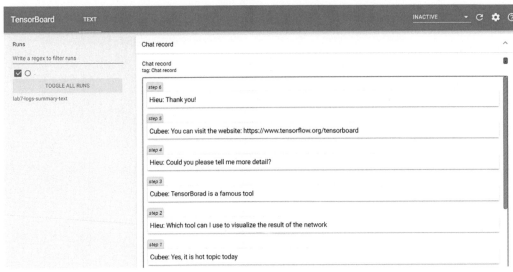

FIG. 7.7   Observing the text conversation written by "tf.summary.text" on the text dashboard.

7.1.4 tf.summary.audio

"tf.summary.audio" is an API of "tf.summary" module used to write an audio summary. Its result can display on the audio dashboard of TensorBoard. The syntax of "tf.summary.audio" is expressed as follows:

> tf.summary.audio(name, data, sample_rate, step=None, max_outputs=3,
> encoding=None,description=None)

Arguments:

- name: a name for the audio summary
- data: a tensor data with shape (n, m, c), where n, m, and c are the number of audios, number of frames, and the number of channels, respectively
- sample_rate: an integer number, representing the sample rate
- step: explicit monotonic step value
- max_outputs: a positive integer number, indicating the number of audios is emitted at each step, the default value is set to three
- encoding: a constant string, representing the type of audio formats such as "wav," "mp3," and so on
- description: optional description for the audio summary

The following program example explains how to use "tf.summary.audio" API. Please download the test audio from the link below and unzip it under the root directory: https://drive.google.com/open?id=1V4nNQ-ZQMBUez ZEFWZoAZ62dORevUGPZ

- Read an audio

```python
#create a function to read audio file
def read_audio(file):
    audio_string = tf.io.read_file(file)  # read file
    audio, fs = tf.audio.decode_wav(audio_string)
    # tf.summary.audio requires the input format to be [n(audio), m(frames), c(channels)]
    # audio data above only has [t(frames), c(channels)], so need to add a dimension
    audio = tf.expand_dims(audio, axis=0)
    return audio, fs
audio, fs = read_audio('./audio/cat.wav')  # Read audio file
```

- Write an audio to the log file and display it on TensorBoard, as shown in Fig. 7.8

```
with summary_writer.as_default():
    tf.summary.audio('cat', audio, fs, step=0)  # Save audio information
```

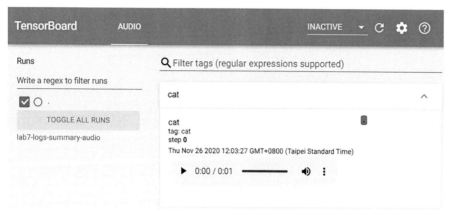

FIG. 7.8   Observing the audio written by "tf.summary.audio" on the audio dashboard.

7.1.5 tf.summary.histogram

"tf.summary.histogram" is an API of "tf.summary" module used to write a histogram summary. Its result can display on the histograms dashboard and distributions dashboard of TensorBoard. The syntax of "tf.summary.histogram" is expressed as follows:

```
tf.summary.histogram(name, data, step=None, buckets=None,
    description=None)
```

Arguments:

- name: a name for the histogram summary
- data: a tensor data
- step: explicit monotonic step value
- buckets: a positive integer number, representing the number of buckets
- description: optional description for the audio summary

Fig. 7.9 shows the weight distribution of a neural network layer obtained by using "tf.keras.callbacks.TensorBoard."

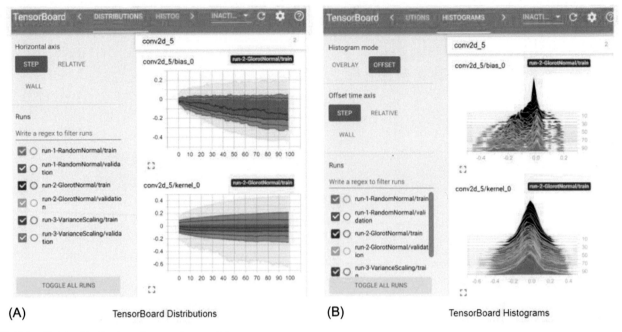

FIG. 7.9   Weight distribution of a neural network layer.

Because the "tf.keras.callbacks.TensorBoard" does not support writing the output distribution of the layers, "tf. summary.histogram" API is recommended if this distribution needs to be visualized. The following program example explains how to use tf.summary.histogram API.

- Write 100 sets of random values to a log file and display them on TensorBoard, as shown in Figs. 7.10 and 7.11.

```
with summary_writer.as_default():
    for i, offset in enumerate(tf.range(0, 10, delta=0.1, dtype=tf.float64)):
        tf.summary.histogram('Normal distribution 2', data+offset, step=i)
```

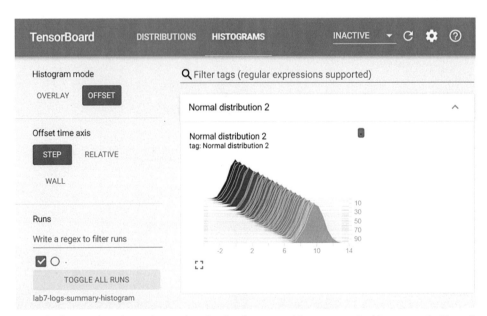

FIG. 7.10   Observing multiple sets of random values written by the tf.summary.histogram on the histograms dashboard.

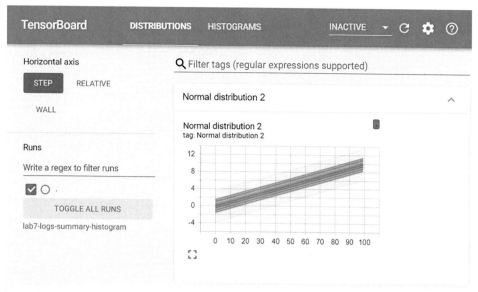

FIG. 7.11 Observing multiple sets of random values written by "tf.summary.histogram" on the distributions dashboard.

7.2 Experiment 1: Using tf.summary.image API to visualize training results

In order to help the reader with a more in-depth understanding of the APIs of the tf.summary module, this experiment employs the "tf.summary.image" API to write the confusion matrix result of a network model trained on the CIFAR-10 dataset and display this result on the images dashboard of TensorBoard. Fig. 7.12 shows the visualization result of the trained network model.

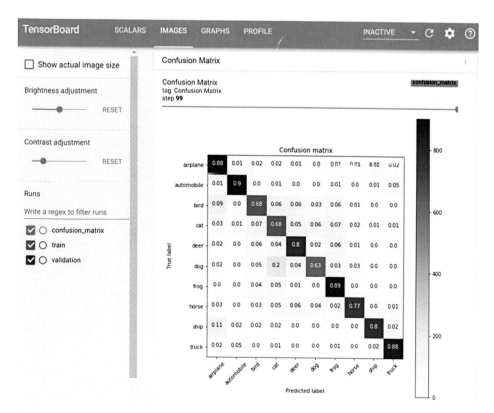

FIG. 7.12 The confusion matrix result on the images dashboard.

Fig. 7.13 is a flowchart of the source code for building a network model and visualizing the output result.

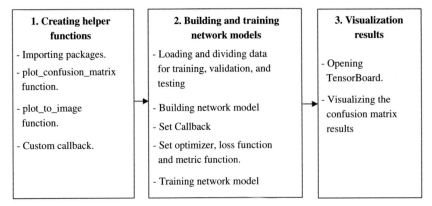

FIG. 7.13   Flowchart of the source code for building the network model and visualizing training result.

1. Creating helper functions

(a) Import necessary packages:

```
import os
import io
import numpy as np
import tensorflow as tf
import tensorflow_datasets as tfds
import matplotlib.pyplot as plt
from tensorflow import keras
from tensorflow.keras import layers
# Import parse_aug_fn and parse_fn functions from the preprocessing.py file
from preprocessing import parse_aug_fn, parse_fn
```

(b) Creating plot_confusion_matrix function.

- Confusion matrix function: use tf.math.confusion_matrix to compute the confusion matrix from expected outputs (y_true) and predictions (y_pred).

```
y_true = [2, 1, 0, 2, 2, 0, 1, 1]
y_pred = [0, 1, 0, 2, 2, 0, 2, 1]
cm = tf.math.confusion_matrix(y_true, y_pred, num_classes=3).numpy()
print(cm)
```

Result: [[2 0 0]
          [0 2 1]
          [1 0 2]]

Supplementary explanation

Fig. 7.14A shows an example of a confusion matrix in which the matrix columns represent the prediction label, and the matrix rows represent the true label. In Fig. 7.14B, the number 2 in the middle indicates that the true label is 1, and the number of correct predictions as 1 is 2. In Fig. 7.14C, the number 2 in the bottom left corner indicates that the true label is 2, and the number of false predictions as 0 is 2.

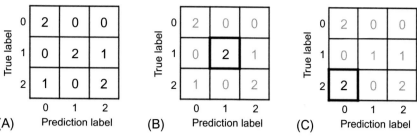

FIG. 7.14 Confusion matrix analysis.

- Create a "plot_confusion_matrix" function: the numbers in the confusion matrix represents the percentage of prediction

```
def plot_confusion_matrix(cm, class_names):
    """
    :param cm (shape = [n, n]): Confusion matrix.
    :param class_names (shape = [n]): category names.
```

```
    """
    # normalization of confusion matrix
    cm = np.around(cm.astype('float') / cm.sum(axis=1)[:, np.newaxis], decimals=2)

    # Set the size of the figure (for displaying)
    figure = plt.figure(figsize=(8, 8))
    # According to the value of "cm", change the color
    plt.imshow(cm, interpolation='nearest', cmap=plt.cm.Blues)
    # The title of the image
    plt.title("Confusion matrix")
    # Scale
    tick_index = np.arange(len(class_names))
    # Y-axis displays category name
    plt.yticks(tick_index, class_names)
        # The x axis displays the category name and rotates the category name by 45 degrees
    plt.xticks(tick_index, class_names, rotation=45)
    # Create a color bar on the right side of the image
    plt.colorbar()

    # Enter the prediction values in each grid cell of Confusion matrix
    threshold = cm.max() / 2.
    for i in range(cm.shape[0]):
        for j in range(cm.shape[1]):
    # If the color of the grid cell is too dark, use white text, otherwise use black text
            color = "white" if cm[i, j] > threshold else "black"
            plt.text(j, i, cm[i, j], horizontalalignment="center", color=color)
    plt.ylabel('True label')
    plt.xlabel('Predicted label')
    # Adjust the position of the image
    plt.tight_layout()
    return figure
img = plot_confusion_matrix(cm, [0, 1, 2])
```

Result:

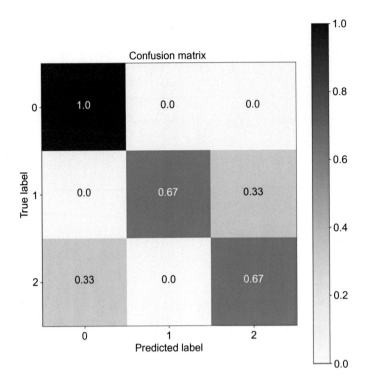

Supplementary explanation

Confusion matrix with normalization: The code below is a normalization process of confusion matrix. Fig. 7.15 shows the conversion operation.

```
# np.around: decimals=2 means that round the value to 2 decimal places
cm = np.around(cm.astype('float') / cm.sum(axis=1)[:, np.newaxis], decimals=2)
```

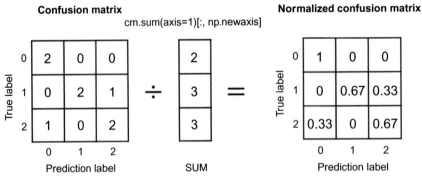

FIG. 7.15   The normalization process of confusion matrix.

(c) Create a "plot_to_image" function

- Convert Matplotlib-style image to TensorFolow-style image for visualizing on TensorBorad.

```python
def plot_to_image(figure):
    """ convert Matplotlib-style image to TensorFolow-style image """
    # Save Matplotlib-style image using PNG format
    buf = io.BytesIO()
    plt.savefig(buf, format='png')
    plt.close(figure)
    buf.seek(0)
    # convert image for using on TensorFlow
    image = tf.image.decode_png(buf.getvalue(), channels=4)
    image = tf.expand_dims(image, 0)
    return image
```

(d) Creating custom callback

After each epoch during the training process, an image of the Confusion matrix is generated for displaying on the TensorBoard.

```python
class ConfusionMatrix(tf.keras.callbacks.Callback):
    def __init__(self, log_dir, test_data, class_name):
        super(ConfusionMatrix, self).__init__()
        self.log_dir = log_dir
        self.test_data = test_data
        self.class_names = class_name
        self.num_classes = len(class_name)

    def on_train_begin(self, logs=None):
        path = os.path.join(self.log_dir, 'confusion_matrix')
        # Create TensorBoard log file
        self.writer = tf.summary.create_file_writer(path)

    def on_epoch_end(self, epoch, logs=None):
        # Calculate Confusion matrix
        total_cm = np.zeros([10, 10])
        for x, y_true in self.test_data:
            y_pred = self.model.predict(x)
            y_pred = np.argmax(y_pred, axis=1)
            y_true = np.argmax(y_true, axis=1)
            cm = tf.math.confusion_matrix(y_true, y_pred,
                        num_classes=self.num_classes).numpy()
            total_cm += cm

        # Plot confusion matrix
        figure = plot_confusion_matrix(total_cm, class_names=self.class_names)
        # Convert Matplotlib-style image for using on TensorFlow
        cm_image = plot_to_image(figure)

        # Write the converted image
        with self.writer.as_default():
            tf.summary.image("Confusion Matrix", cm_image, step=epoch)
```

2. Building and training network model

(a) Loading and dividing data

- Load CIFAR-10 dataset:

```
#Divide training data: 9 part for training, 1 part for validation
train_split, valid_split = ['train[:90%]', 'train[90%:]']
# # get the training data
train_data, info = tfds.load("cifar10", split=train_split, with_info=True)
# get the valid data
valid_data = tfds.load("cifar10", split=valid_split)
# get test data
test_data = tfds.load("cifar10", split=tfds.Split.TEST)
# Get the category name of the CIFAR-10 dataset
class_name = info.features['label'].names
```

- Dataset setting:

```
AUTOTUNE = tf.data.experimental.AUTOTUNE  # Automatic adjustment mode
batch_size = 64  # Batch size
train_num = int(info.splits['train'].num_examples / 10) * 9  # Number of training data
train_data = train_data.shuffle(train_num)  # Shuffle the training data
# Training data
train_data = train_data.map(map_func=parse_aug_fn, num_parallel_calls=AUTOTUNE)
# Set batch size and turn on prefetch mode
train_data = train_data.batch(batch_size).prefetch(buffer_size=AUTOTUNE)


# Validation data
valid_data = valid_data.map(map_func=parse_fn, num_parallel_calls=AUTOTUNE)
# Set batch size and turn on prefetch mode
valid_data = valid_data.batch(batch_size).prefetch(buffer_size=AUTOTUNE)


# Test data
test_data = test_data.map(map_func=parse_fn, num_parallel_calls=AUTOTUNE)
#Set batch size and turn on prefetch mode
test_data = test_data.batch(batch_size).prefetch(buffer_size=AUTOTUNE)
```

(b) Building a network model

Table 7.1 shows the architecture of the network model.

TABLE 7.1   The architecture of the network model.

Name	Architecture	Description
Model-1	- Input layer with shape of (32,32,3) - Five convolutional layers, followed by rectified linear unit (ReLu) activation functions - One max pooling layer - One flatten layer for flattening the input into a one-dimensional Tensor - One fully connected layer - One dropout layer with a discard rate of 50% - Output fully connected layer with 10 neurons	- Using high-level Keras API to build and train the network model - Using custom Callback to write image result for displaying on TensorBoard

- Build network model

```
inputs = keras.Input(shape=(32, 32, 3))
x = layers.Conv2D(64, 3, activation='relu')(inputs)
x = layers.MaxPool2D()(x)
x = layers.Conv2D(128, 3, activation='relu')(x)
x = layers.Conv2D(256, 3, activation='relu')(x)
x = layers.Conv2D(128, 3, activation='relu')(x)
x = layers.Conv2D(64, 3, activation='relu')(x)
x = layers.Flatten()(x)
x = layers.Dense(64, activation='relu')(x)
x = layers.Dropout(0.5)(x)
outputs = layers.Dense(10)(x)
# Create a network model
model_1 = keras.Model(inputs, outputs, name='model 1')
model_1.summary()
```

Result:

```
Model: "model 1"

Layer (type)                    Output Shape              Param #
=================================================================
input_1 (InputLayer)            [(None, 32, 32, 3)]       0

conv2d (Conv2D)                 (None, 30, 30, 64)        1792

max_pooling2d (MaxPooling2D)    (None, 15, 15, 64)        0

conv2d_1 (Conv2D)               (None, 13, 13, 128)       73856

conv2d_2 (Conv2D)               (None, 11, 11, 256)       295168

conv2d_3 (Conv2D)               (None, 9, 9, 128)         295040

conv2d_4 (Conv2D)               (None, 7, 7, 64)          73792

flatten (Flatten)               (None, 3136)              0

dense (Dense)                   (None, 64)                200768

dropout (Dropout)               (None, 64)                0

dense_1 (Dense)                 (None, 10)                650
=================================================================
Total params: 941,066
Trainable params: 941,066
Non-trainable params: 0
```

(c) Set Callback function

```
# Save training log
logs_dirs = 'lab7-logs-images'
model_cbk = keras.callbacks.TensorBoard(logs_dirs)
# Save Confusion matrix image
save_cm = ConfusionMatrix(logs_dirs, test_data, class_name)
```

(d) Set the optimizer, loss function, and metric function for training

```
model_1.compile(keras.optimizers.Adam(),
        loss=keras.losses.CategoricalCrossentropy(from_logits=True),
        metrics=[keras.metrics.CategoricalAccuracy()])
```

(e) Training network model

```
model_1.fit(train_data,
        epochs=100,
        validation_data=valid_data,
        callbacks=[model_cbk, save_cm])
```

Result:

```
Epoch 96/100
704/704 [==============================] - 13s 18ms/step - loss: 0.7863 - categorical_accuracy: 0.7395 - val_loss: 0.6702
 - val_categorical_accuracy: 0.7900
Epoch 97/100
704/704 [==============================] - 13s 18ms/step - loss: 0.7890 - categorical_accuracy: 0.7370 - val_loss: 0.6618
 - val_categorical_accuracy: 0.7960
Epoch 98/100
704/704 [==============================] - 13s 19ms/step - loss: 0.7761 - categorical_accuracy: 0.7423 - val_loss: 0.6465
 - val_categorical_accuracy: 0.7918
Epoch 99/100
704/704 [==============================] - 13s 18ms/step - loss: 0.7992 - categorical_accuracy: 0.7354 - val_loss: 0.6544
 - val_categorical_accuracy: 0.7866
Epoch 100/100
704/704 [==============================] - 13s 18ms/step - loss: 0.7820 - categorical_accuracy: 0.7400 - val_loss: 0.6489
 - val_categorical_accuracy: 0.7966

<tensorflow.python.keras.callbacks.History at 0x7f28b0183da0>
```

3. Visualization of confusion matrix results on TensorBoard
   - Open TensorBoard through command line to view training records

```
tensorboard --logdir lab7-logs-images
```

   - Use TensorBoard to visualize the confusion matrix results

The confusion matrix result of the network model in each epoch can be observed through the TensorBoard. The progress bar is used to observe the prediction results of the network model in different epochs, as shown in Fig. 7.16.

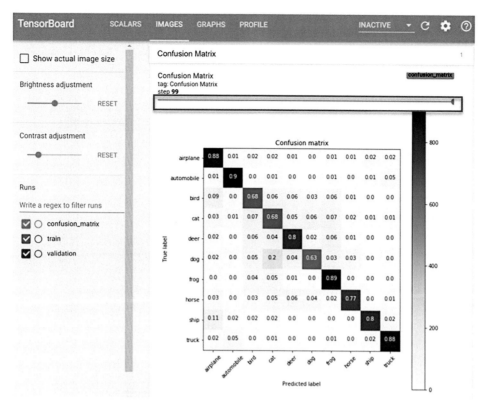

FIG. 7.16 Observing the confusion matrix result on the images dashboard.

From confusion matrix result in Fig. 7.17, the larger the values in the highlight boxes, the better the performance of the network model.

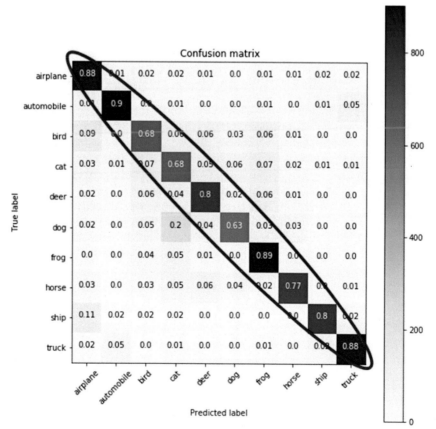

FIG. 7.17 The confusion matrix result (1).

In addition, the relationship between classes can also be observed from the confusion matrix. As shown in Fig. 7.18, there is a 20% prediction of a "cat" class as a "dog" class, which means the network model easily confuses "cat" class with "dog" class.

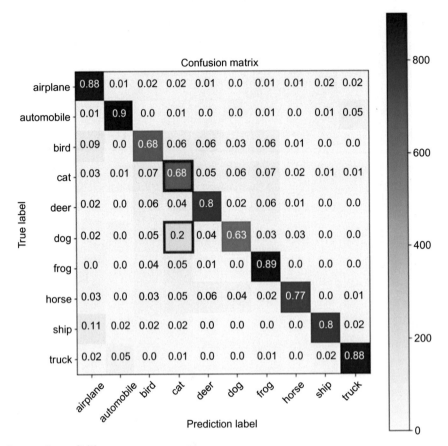

FIG. 7.18   The confusion matrix result (2).

7.3 Experiment 2: Hyperparameter tuning with TensorBoard HParams

There are many issues in developing a good neural network model for a specific problem, such as how many network layers should be used, which type of optimizer is better, how to set a suitable learning rate value, whether to apply dropout technique or batch normalization method, and so on. These adjustable parameters are collectively referred to as hyperparameters and they directly affect the training results. Therefore, an important step in developing neural network models is to choose the best hyperparameters. The last section of this chapter introduces a hyperparameter tuning technique using the HParams dashboard provided by TensorFlow, which can help to identify the most suitable hyperparameters for a specific problem. This technique is explained directly through a program example that builds 36 network models with different combinations of hyperparameters that are trained on the CIFAR-10 dataset for analysis and comparison of obtained results. The hyperparameters used for building neural network models are as follows:

- Image augmentation (IA): Yes or No
- Batch normalization (BN): Yes or No
- Learning rate (LR): 0.001, 0.01, or 0.03
- Weight initialization method: Random Normal, Glorot Normal, or He Normal

7.3.1 HPARAMS dashboard on TensorBoard

- Open TensorBoard through command line:

```
tensorboard --logdir lab7-logs-hparams
```

- Open HParams dashboard: After opening the TensorBoard, click on HParams in the list options, as shown in Fig. 7.19.

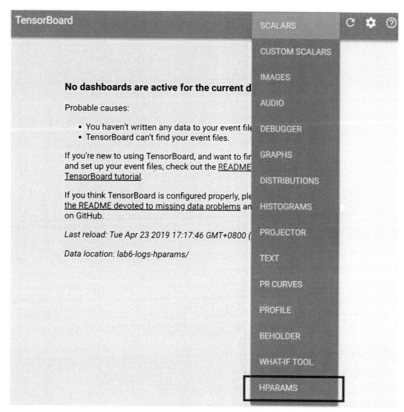

FIG. 7.19 List options in TensorBoard.

- Viewing HParams dashboard

After clicking on HParams, a dashboard is opened, as shown in Fig. 7.20. The left pane of the HParams dashboard shows the views of hyperparameters, metrics values, run status, sorting information in the table view, and number of session groups. The main dashboard displays three different views including the list of runs on the "table view", each run for each hyperparameter and metric on the "parallel coordinates view", and plots for comparison of each hyperparameter with each metric on the "scatter plot view". In Fig. 7.20, because there are not any data written, there is not information displayed in the HParams dashboard.

FIG. 7.20   HParams dashboard in TensorBoard.

7.3.2 Code examples

Fig. 7.21 is a flowchart of the source code for building and analyzing 36 neural network models with various hyperparameters.

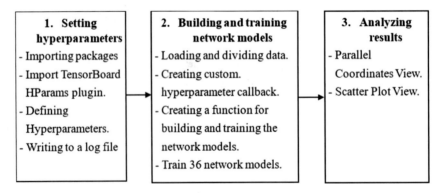

FIG. 7.21   Flowchart of the source code for building and analyzing 36 neural networks with different combinations of hyperprameters.

1. Setting hyperparameters

    (a) Import packages

```
        import tensorflow as tf
import tensorflow_datasets as tfds
from tensorflow import keras
from tensorflow.keras import layers
from tensorflow.keras import initializers
# Import parse_aug_fn and parse_fn functions from the preprocessing.py file
from preprocessing import parse_aug_fn, parse_fn
```

    (b) Import TensorBoard HParams plugin

```
from tensorboard.plugins.hparams import api as hp
```

    (c) Defining Hyperparameters

The hyperparameters used for building network models are listed as below:

- Imgae Augmentation: Yes or No.
- Batch Normalization (BN): Yes or No.
- Learning rate (LR): 0.001, 0.01, 0.03.
- Weight initialization method: Random Normal, Glorot Normal, He Normal.

```
hparam_ia = hp.HParam('Imgae_Augmentation', hp.Discrete([False, True]))
hparam_bn = hp.HParam('Batch_Normalization', hp.Discrete([False, True]))
hparam_init = hp.HParam('Weight_Initialization', hp.Discrete(['RandomNormal_0.01std',
                                                              'glorot_normal',
                                                              'he_normal']))
hparam_lr = hp.HParam('Learning_Rate', hp.Discrete([0.001, 0.01, 0.03]))
```

    (d) Writing to a log file

- Write the experimental hyperparameter information and metric information to the TensorBoard log file:

```
# Create TensorBoard logs file
logs_dirs = os.path.join('lab7-logs-hparams', 'hparam_tuning')
root_logdir_writer = tf.summary.create_file_writer(logs_dirs)
with root_logdir_writer.as_default():
    hp.hparams_config(hparams=[hp_ia, hp_bn, hp_init, hp_lr], metrics=[hp_metric])

# Label information
metric = 'Accuracy'
log_dirs = "lab7-logs-hparams/hparam_tuning"

with tf.summary.create_file_writer(log_dirs).as_default():
    # Write hyperparameter information and metric information to the log file
    hp.hparams_config(
        hparams=[hparam_ia, hparam_bn, hparam_init, hparam_lr],
        metrics=[hp.Metric(metric, display_name='Accuracy')],
    )
```

After executing the program above, the TensorBoard screen has more information of hyperparameters and metric values, as shown in Fig. 7.22.

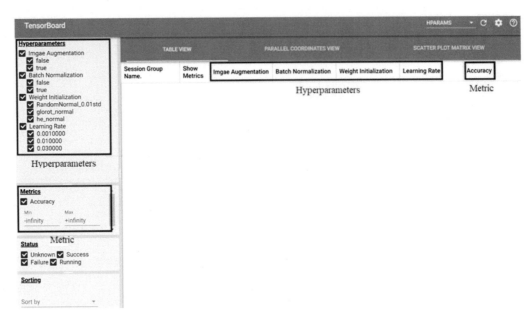

FIG. 7.22   TensorBoard HParamas with hyperparameters and metric information.

2. Building and training network models

(a) Loading and dividing data

- Load CIFAR-10 dataset:

```
# Get data
valid_split, train_split = tfds.Split.TRAIN.subsplit([10, 90])
# training data without image augmentation
train_data_noaug, info = tfds.load("cifar10", split=train_split, with_info=True)
# training data with image augmentation
train_data_aug = tfds.load("cifar10", split=train_split)
# validation data
valid_data = tfds.load("cifar10", split=valid_split)
```

- Prepare training data: with image augmentation and without image augmentation

```
AUTOTUNE = tf.data.experimental.AUTOTUNE  # Automatic adjustment mode
batch_size = 64  # batch size
train_num = int(info.splits['train'].num_examples / 10) * 9


# Set the training data without image augmentation
train_data_noaug = train_data_noaug.shuffle(train_num)  # shuffle training data
train_data_noaug=train_data_noaug.map(map_func=parse_fn,
                    num_parallel_calls=AUTOTUNE)
# Set batch size and turn on prefetch mode
train_data_noaug=train_data_noaug.batch(batch_size).prefetch(buffer_size=AUTOTUNE)


# training data with image augmentation
train_data_aug = train_data_aug.shuffle(train_num)  # Shuffle data
train_data_aug=train_data_aug.map(map_func=parse_aug_fn,
                    num_parallel_calls=AUTOTUNE)
# Set batch size and turn on prefetch mode
train_data_aug = train_data_aug.batch(batch_size).prefetch(buffer_size=AUTOTUNE)


# Validation data
valid_data = valid_data.map(map_func=parse_fn, num_parallel_calls=AUTOTUNE)
# Set batch size and turn on prefetch mode
valid_data = valid_data.batch(batch_size).prefetch(buffer_size=AUTOTUNE)
```

(b) Creating custom hyperparameter callback function

The custom hyperparameter callback function is created to write the hyperparameters and metric values for displaying on TensorBoard.

```
class HyperparameterCallback(tf.keras.callbacks.Callback):
    # Call when the category is created
    def __init__(self, log_dir, hparams):
        super(HyperparameterCallback, self).__init__()
        self.log_dir = log_dir
```

```
        self.hparams = hparams
        self.best_accuracy = 0
        self.writer = None


    # call before starting training to create a log file
    def on_train_begin(self, logs=None):
        self.writer = tf.summary.create_file_writer(self.log_dir)


    # after each Epoch, if the model progresses, #its weights is updated.
    def on_epoch_end(self, epoch, logs=None):
        current_accuracy = logs.get('val_categorical_accuracy')
        if current_accuracy > self.best_accuracy:
            self.best_accuracy = current_accuracy


    # Call at the end of the training, and save the training hyperparameters and the best
    # accuracy in the log file.
    def on_train_end(self, logs=None):
        with self.writer.as_default():
            hp.hparams(self.hparams)  # write the weight parameters of this training
            tf.summary.scalar(metric, self.best_accuracy, step=0)
```

(c) Creating a function for building and training the network models.

The network layers used in the models are as follows:

- keras.Input: input layer with the size of $32 \times 32 \times 3$
- layers.Conv2D: convolutional layer (weight initialization: Random normal, Glorot normal, or He normal)
- layers.BatchNormalization: batch normalization layer with preset parameters (hyperparameters: use or do not use)
- layers.ReLU: ReLU activation function layer
- layers.MaxPool2D: pooling layer, for downsampling the feature map
- layers.Flatten: flatten layer, for flattening the input into a one-dimensional Tensor
- layers.Dropout: dropout layer, which randomly sets input units to 0 with a frequency of 50 at each step during training time
- layers.Dense: fully connected layer; if this layer is used as the hidden layer of the network, the ReLU activation function will be adopted. ReLU activation function will be replaced with softmax activation function if this layer is used as the output layer of the network

```python
def train_test_model(logs_dir, hparams):
    """

    logs_dir: the location of the currently executed task log file

    hparams: Incoming hyperparameter
    """
    # Weight initialization: using normal distribution (std 0.01), Glorot, or He method
    if hparams[hparam_init] == "glorot_normal":
        init = initializers.glorot_normal()
    elif hparams[hparam_init] == "he_normal":
        init = initializers.he_normal()
    else:
        init = initializers.RandomNormal(0, 0.01)


    inputs = keras.Input(shape=(32, 32, 3))
    x = layers.Conv2D(64, (3, 3), kernel_initializer=init)(inputs)
        # Choose "add" or "not add" Batch Normalization layer
    if hparams[hparam_bn]: x = layers.BatchNormalization()(x)
    x = layers.ReLU()(x)
    x = layers.MaxPool2D()(x)
    x = layers.Conv2D(128, (3, 3), kernel_initializer=init)(x)
    # choose "add" or "not add" Batch Normalization layer
    if hparams[hparam_bn]: x = layers.BatchNormalization()(x)
    x = layers.ReLU()(x)
    x = layers.Conv2D(256, (3, 3), kernel_initializer=init)(x)
    # choose "add" or "not add" Batch Normalization layer
    if hparams[hparam_bn]: x = layers.BatchNormalization()(x)
    x = layers.ReLU()(x)
    x = layers.Conv2D(128, (3, 3), kernel_initializer=init)(x)
    # choose "add" or "not add" Batch Normalization layer
    if hparams[hparam_bn]: x = layers.BatchNormalization()(x)
    x = layers.ReLU()(x)
    x = layers.Conv2D(64, (3, 3), kernel_initializer=init)(x)
    # choose "add" or "not add" Batch Normalization layer
    if hparams[hparam_bn]: x = layers.BatchNormalization()(x)
    x = layers.ReLU()(x)
    x = layers.Flatten()(x)
    x = layers.Dense(64, kernel_initializer=init)(x)
```

```python
# choose "add" or "not add" Batch Normalization layer
if hparams[hparam_bn]: x = layers.BatchNormalization()(x)
x = layers.ReLU()(x)
x = layers.Dropout(0.5)(x)
outputs = layers.Dense(10, activation='softmax')(x)
# Create a network model
model = keras.Model(inputs, outputs, name='model')


# Save training log
model_tb = keras.callbacks.TensorBoard(log_dir=logs_dir, write_graph=False)
# Save the best model's weight
model_mckp = keras.callbacks.ModelCheckpoint(logs_dir +'/best-model.hdf5',
                    monitor='val_categorical_accuracy',
                    save_best_only=True,
                    mode='max')
    # Set the conditions for early stopping
model_els = keras.callbacks.EarlyStopping(monitor='val_categorical_accuracy' ,
                    min_delta=0,
                    patience=30,
                    mode='max')
    # Custom callback to write the hyperparameters and metric (accuracy) of the
    #training model
model_hparam = HyperparameterCallback(logs_dir + 'hparam_tuning', hparams)


# Set the optimizer, loss function, and metric function
# The learning rate is: 0.001, 0.01 or 0.03
model.compile(keras.optimizers.Adam(hparams[hparam_lr]),
        loss=keras.losses.CategoricalCrossentropy(),
        metrics=[keras.metrics.CategoricalAccuracy()])
# Hyperparameters: use or do not use "image augmentation" to train the network
if hparams[hparam_ia]:
   history = model.fit(train_data_aug,
                epochs=100,
                validation_data=valid_data,
                callbacks=[model_tb,model_mckp,model_els,model_hparam])
else:
   history = model.fit(train_data_noaug,
                epochs=100,
                validation_data=valid_data,
                callbacks=[model_tb,model_mckp,model_els,model_hparam])
```

(d) Training 36 network models with different combinations of hyperparameters
- Table 7.2 shows the combinations of hyperparameters for training 36 neural network models.

TABLE 7.2   Combinations of hyperparameters for training neural network models.

Name	Image augmentation		Batch normalization		Learning rate			Weight initialization		
	Yes	No	Yes	No	0.001	0.01	0.03	Random normal	Glorot normal	He normal
Model_1	✓		✓		✓			✓		
Model_2	✓		✓		✓				✓	
Model_3	✓		✓		✓					✓
Model_4	✓		✓			✓		✓		
Model_5	✓		✓			✓			✓	
Model_6	✓		✓			✓				✓
Model_7	✓		✓				✓	✓		
Model_8	✓		✓				✓		✓	
Model_9	✓		✓				✓			✓
Model_10	✓			✓	✓			✓		
Model_11	✓			✓	✓				✓	
Model_12	✓			✓	✓					✓
Model_13	✓			✓		✓		✓		
Model_14	✓			✓		✓			✓	
Model_15	✓			✓		✓				✓
Model_16	✓			✓			✓	✓		
Model_17	✓			✓			✓		✓	
Model_18	✓			✓			✓			✓
Model_19		✓	✓		✓			✓		
Model_20		✓	✓		✓				✓	
Model_21		✓	✓		✓					✓
Model_22		✓	✓			✓		✓		
Model_23		✓	✓			✓			✓	
Model_24		✓	✓			✓				✓
Model_25		✓	✓				✓	✓		
Model_26		✓	✓				✓		✓	
Model_27		✓	✓				✓			✓
Model_28		✓		✓	✓			✓		
Model_29		✓		✓	✓				✓	
Model_30		✓		✓	✓					✓
Model_31		✓		✓		✓		✓		
Model_32		✓		✓		✓			✓	
Model_33		✓		✓		✓				✓
Model_34		✓		✓			✓	✓		
Model_35		✓		✓			✓		✓	
Model_36		✓		✓			✓			✓

- Training network models

```
session_id = 1  # the id of Training task
# the place to store log files
logs_dir = os.path.join('lab7-logs-hparams', 'run-{}')
for ia in hparam_ia.domain.values:
    for bn in hparam_bn.domain.values:
        for init in hparam_init.domain.values:
            for lr in hparam_lr.domain.values:
                # Display the current training task id
                print('--- Running training session {}'.format(session_id))
                # Set the hyperparameters for this training
                hparams={hparam_ia:ia, hparam_bn:bn, hparam_init:init, hparam_lr:lr}
                # the place to store log files
                logs_dir=os.path.join("lab7-logs-hparams","run-{}".format(session_id))
                # Create, compile and train network models
                train_test_model(logs_dir, hparams)
                session_id += 1  # id+1
```

3. Analyzing the training results through TensorBoard HParams.

The hyperparameters and training results of 36 neural network models are displayed on the HParams dashboard, as shown in Fig. 7.23.

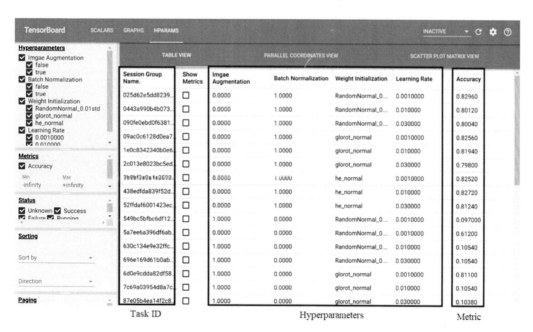

FIG. 7.23   TensorBoard HParams with 36 trained network models.

(a) Parallel Coordinates View

To observe the results with parallel coordinates view, click "PARALLEL COORDINATES VIEW" in the dashboard, as shown in Fig. 7.24.

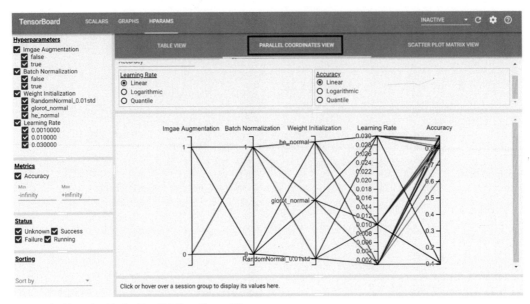

FIG. 7.24   HParams dashboard with the Parallel Coordinates View.

From the results on the parallel coordinates view, as shown in Fig. 7.25, the network models with high accuracy employed image augmentation and batch normalization techniques, whereas using the different types of weight initialization methods or learning rates has little effect on training results.

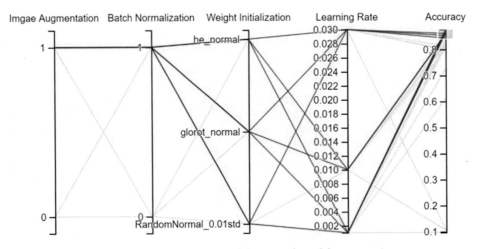

FIG. 7.25   The impact of different hyperparameters on the accuracy of the network models.

In TensorBoard HParams, the range of accuracy values can be set to filter unnecessary information, as shown in Fig. 7.26. From the HParams dashboard, it can be observed that if batch normalization is applied, the choice of weight initialization methods is less careful because it does not change much about the accuracy of the models.

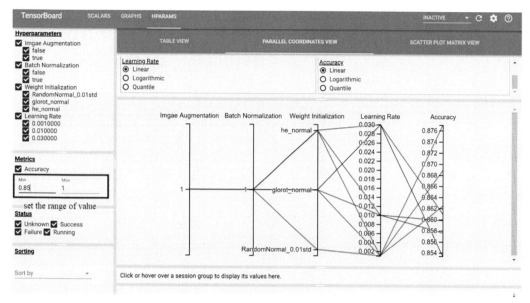

FIG. 7.26 TensorBoard HParams with the limitation of the accuracy range.

We can also display the results of the models with a specific learning rate value, such as displaying the result of trained models with the learning rate value of 0.001, as shown in Fig. 7.27. As shown, the three network models with the highest accuracy are all trained using a learning rate of 0.001.

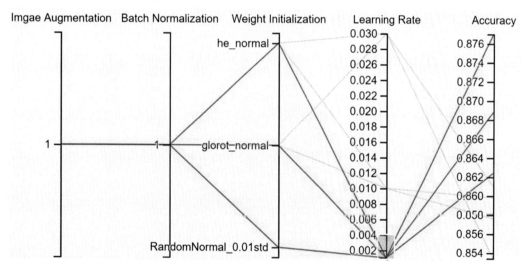

FIG. 7.27 TensorBoard HParams with the selection of running a specific learning rate.

Supplement explanation

Results in Fig. 7.27 show that the network model achieves the best accuracy when trained with a learning rate of 0.001. However, the comparison is not fair because the convergence rate of the network model using a learning rate of 0.001 is much slower than that of using a learning rate of 0.01 or 0.03. Furthermore, using callback functions such as ReduceLROnPlateau also reduces this value, so it is recommended to use the higher learning rate first, and then reduce this value gradually during the training process.

(b) Scatter Plot View

To display plots comparing each hyperparameter with each metric, click "SCATTER PLOT VIEW" in the dashboard, as shown in Fig. 7.28. We leave this to the readers to explore on their own. The analysis method is similar to the "Parallel Coordinates View" subsection.

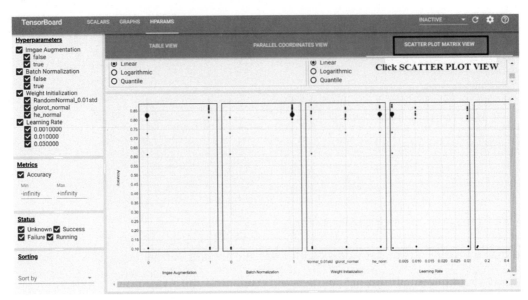

FIG. 7.28   HParams dashboard with Scatter Plot View.

8

Convolutional neural network architectures

8.1 Popular convolutional neural network architectures

This chapter introduces some of the most representative architectures of convolutional neural networks (CNNs), including LeNet [1], AlexNet [2], VGG [3], GoogLeNet [4], and ResNet [5]. These CNNs have been widely applied to many computer vision applications such as image classification, face recognition, object detection, and so on.

8.1.1 LeNet

Yann et al. proposed the LeNet [1] in 1998 for handwritten character recognition. The architecture of LeNet is straightforward and simple, consisting of three convolutional layers named C1, C3, and C5, two subsampling layers named S2 and S4, and one fully connected layer named F6 and Output, as shown in Fig. 8.1.

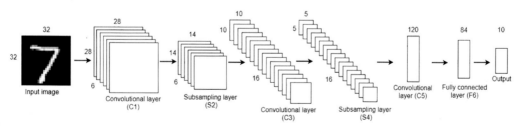

FIG. 8.1 The architecture of LeNet.

- Layer C1: The input grayscale image of LeNet with size of 32×32 is passed through a convolutional layer (C1) with six kernels having size of 5×5 to output six feature maps of size 28×28. C1 has 22,304 connections and 156 trainable parameters.
- Layer S2: The output of C1 is sent to a subsampling layer (S2) with a kernel size of 2×2 and stride of 2, followed by a sigmoid activation function, resulting in six feature maps with a size of 14×14. S2 has 5880 connections and 12 trainable parameters.

- Layer C3: A convolutional layer with 16 kernels having a size of 5×5 receives the output of S2 as input and results in 16 feature maps of size 28×28. C3 has 156,000 connections and 1516 trainable parameters.
- Layer S4: A subsampling layer is the same as the second layer (S2). This layer uses 16 kernels of size 2×2 and a stride of 2 to output 16 feature maps of size 5×5. S4 has 2000 connections and 32 trainable parameters.
- Layer C5: A convolutional layer with 120 kernels of size 5×5 that receives 16 output feature maps with a size 5×5 of S4 as the input and produces 120 feature maps with size of 1×1. C5 has 48,120 trainable parameters.
- Layer F6: A fully connected layer with 84 neurons followed by sigmoid activation. F6 contains 10,164 trainable parameters.
- Output layer: A fully connected layer with 10 neurons, one neuron for each class.

8.1.2 AlexNet

AlexNet [2] was the winner of the 2012 ImageNet Large Scale Visual Recognition Challenge (ILSVRC 2012). It has an architecture that is larger and deeper than that of LeNet [1], as shown in Fig. 8.2. Specifically, AlexNet architecture is composed of eight layers including five convolutional layers with kernels of size 11×11 for the first layer, 5×5 for the second layer, and 3×3 for the other layers, followed by three fully connected layers. The rectified linear unit (ReLU) activation function is adopted at the output of every convolutional layer and fully connected layer, except for the last fully connected layer, which sends its output to the softmax function for final classification. To achieve impressive performance, AlexNet employed dropout and data-augmentation techniques to prevent overfitting during the training process. If LeNet paved the way for CNNs, then AlexNet is important for driving the CNN boom.

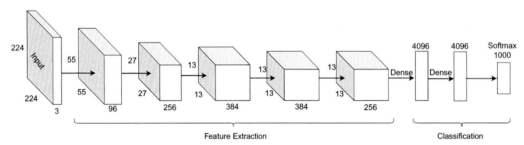

FIG. 8.2   The architecture of AlexNet.

8.1.3 VGG

VGG [3] is a deep CNN for large-scale image recognition introduced by Karen et al. in 2014. VGG has many versions such as VGG-11, VGG-13, VGG-16, and VGG-19, and the main difference between each version is the number of layers. The architecture of VGG-16 consists of 16 convolutional layers, as shown in Fig. 8.3.

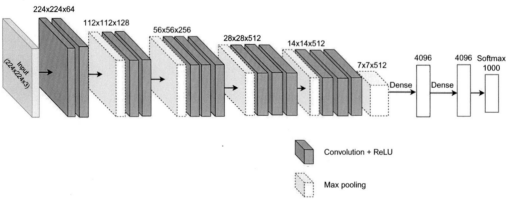

FIG. 8.3   The architecture of VGG-16.

In contrast to previous CNNs, VGG is only composed of convolutional layers with kernels of size 3×3 instead of $5 \times 5, 7 \times 7$, or 9×9. The main reason for this is that a stack of multiple 3×3 convolutional layers can achieve a similar perception field of view as using a $5 \times 5, 7 \times 7$, or 9×9 convolutional layer, while the number of parameters is greatly reduced. Fig. 8.4 shows a stack of two 3×3 convolutional layers, and a single 5×5 convolutional layer. Both have the same input size and output size, but using two 3×3 convolutional layers, only 18 ($3^2 + 3^2$) parameters are required, whereas a single 5×5 convolutional layer would require 25 (5^2) parameters. Note that the channel was omitted for simple computation.

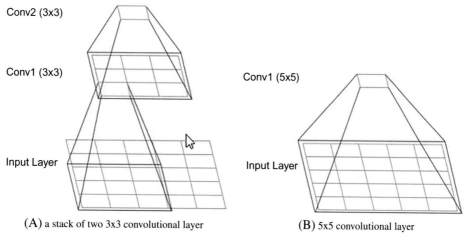

(A) a stack of two 3x3 convolutional layer   (B) 5x5 convolutional layer

FIG. 8.4   A stack of multiple convolutional layers and a single convolutional layer.

8.1.4 GoogLeNet

GoogLeNet [4] is a deep network architecture proposed by Google, also known as Inception-v1, which won the 2014 ILSVRC. The architecture of GoogLeNet is composed of nine Inception Blocks with a depth of 22 layers with parameters, as shown in Fig. 8.5.

The main idea of using Inception Blocks in GoogLeNet is to increase the depth and width of the network while keeping the computation constant. The original inception architecture, also known as naïve Inception architecture, is shown in Fig. 8.6. It contains four parallel layers including 1×1 convolution, 3×3 convolution, 5×5 convolution, and 3×3 max pooling layers. The single output of this module is formed by concatenating all those layers, and it is used as the input of the next stage in the network model. Although the network composed of naïve Inception Blocks can obtain high performance, it requires a lot of parameters and computation.

The Inception Block in GoogLeNet is designed by connecting the 1×1 convolutional layer to 3×3 and 5×5 convolutional layers, and the 3×3 max pooling layer of the naïve Inception version, as shown in Fig. 8.7. To show the benefit of using 1×1 convolutional layers in the Inception Block, let us use the example of computing the number of parameters when using convolutional layers to convert an input with a size of $36 \times 36 \times 128$ to $36 \times 36 \times 64$.

- The first method: The input is directly fed into a 3×3 convolutional layer to generate the expected output. The number of parameters in the convolutional layer is 73,728, as shown in Fig. 8.8.
- The second method: The input is first passed through a 1×1 convolutional layer to reduce the number of feature maps to 16, then the resulting feature maps are sent to a 3×3 convolutional layer to achieve the expected output. The number of parameters of two convolutional layers is 11,264, which is less than the 73,728 parameters in the first method without applying the 1×1 convolutional layer, as shown in Fig. 8.9.

Although the depth of GoogLeNet is 22 layers, by using Inception Blocks the number of parameters of GoogLeNet is still 12 times less than that of AlexNet with 8 layers.

After Google proposed the GoogLeNet architecture, many more versions of the Inception models were introduced. The following is a summary of each version:

- GoogLeNet, also known as Inception-v1 [4]: Published in 2014, it is the first multi-branch architecture that introduced 1×1 convolution to reduce the amount of network computation

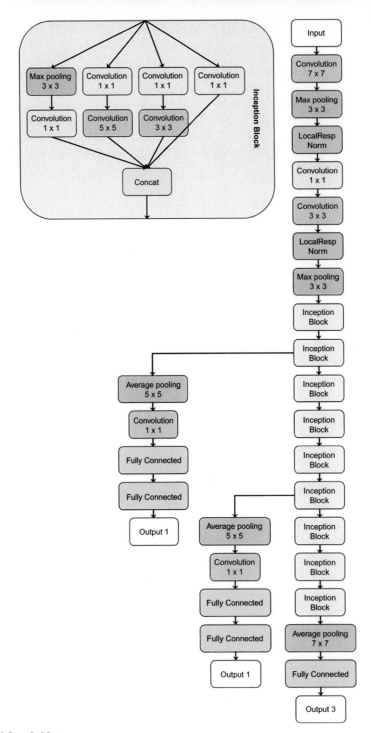

FIG. 8.5 The architecture of GoogLeNet.

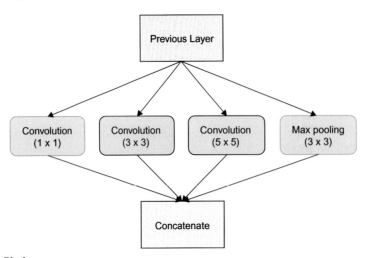

FIG. 8.6 The naïve Inception Block.

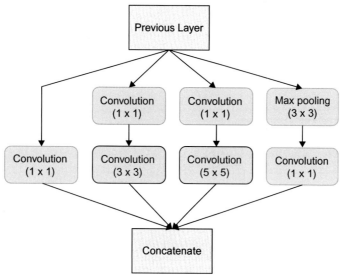

FIG. 8.7 Inception Block in GoogLeNet.

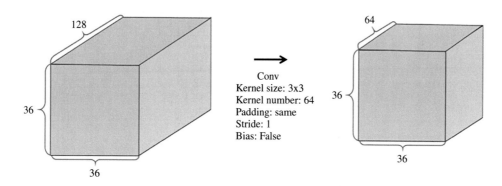

The number of parameters = $(128 \times 3 \times 3 + 10) \times 64 = 73,728$

FIG. 8.8 Computation of the number of learnable parameters in a 3×3 convolutional layer.

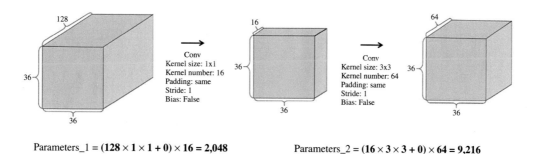

Parameters_1 = $(128 \times 1 \times 1 + 0) \times 16 = 2,048$ Parameters_2 = $(16 \times 3 \times 3 + 0) \times 64 = 9,216$

Total number of parameters = $2,048 + 9,216 = 11,264$

FIG. 8.9 Computation of the number of learnable parameters in a combined architecture of 1×1 and 3×3 convolutional layers.

- Inception-v2 [6]: Published in 2015, introducing batch normalization for speeding up the training and improving the performance of CNNs
- Inception-v3 [7]: Published in 2016, exploring $1 \times n$ and $n \times 1$ convolutions in constructing the Inception module
- Inception-v4 [8]: Published in 2017, using Inception architecture with residual connections
- Xception [9]: Published in 2017, introducing depthwise separable convolution for building CNN architecture instead of Inception modules

8.1.5 ResNet

Microsoft introduced ResNet [5] (Residual Nets) in 2015, when it won that year's ILSVRC. There are five versions of ResNet architecture for ImageNet, including ResNet-18, ResNet-34, ResNet-50, ResNet-101, and ResNet-152. Here, the x in ResNet-x is the number of layers of the corresponding version of ResNet. Fig. 8.10 shows the architecture of ResNet-34.

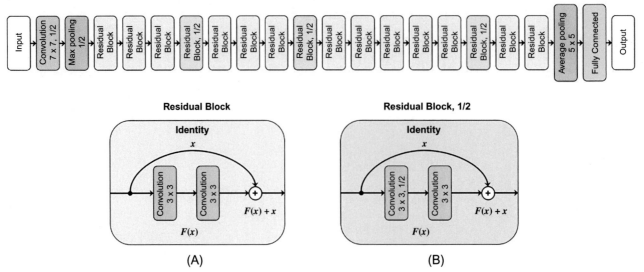

FIG. 8.10   The architecture of ResNet-34. (A) Residual Block: a stack of two 3×3 convolutional layers, where the number of kernels of layers is equal. (B) Residual Block, 1/2: a stack of two 3×3 convolutional layers, where the number of kernels of the first layer is one a half that of the second layer.

In deep learning, the deeper neural networks are able to learn more complex features and achieve better performance than that of the shallower neural networks. However, degradation, which refers to a problem of the accuracy of the neural network, becomes saturated and then degrades quickly during the training process, leading to greater training error [10]. For example, Fig. 8.11 shows the experimental results of two CNNs [5], including a 20-layer neural network and a 56-layer neural network on the CIFAR-10 dataset. As shown, the 20-layer neural network has fewer training and test errors than the 56-layer network.

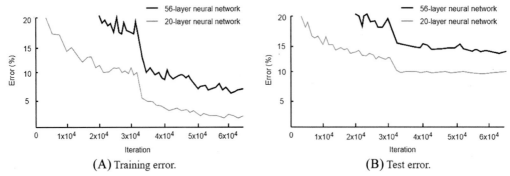

FIG. 8.11   Training and test errors of 20-layer neural network and 56-layer neural network on the CIFAR-10 dataset [5].

To overcome the degradation problem, ResNet uses Residual Blocks for building a deep neural network. Fig. 8.12 shows a typical Residual.

Residual Block

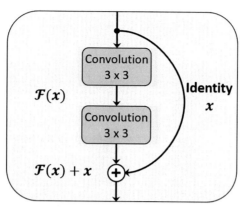

FIG. 8.12   Residual Block, adding a shortcut connection to a pair of 3×3 convolutional layer.

Fig. 8.13 presents the experimental result of training ResNet on the ImageNet dataset. Fig. 8.13A shows the results of an 18-layer network and 34-layer network without using shortcut connections. As shown, the training and validation errors of the 34-layer network are greater than those of the 18-layer network. When adding shortcut connections, that is, using Residual Blocks to construct networks, the training and validation errors of the 34-layer network or ResNet-34 are fewer than those of the 18-layer network or ResNet-18, as shown in Fig. 8.13B.

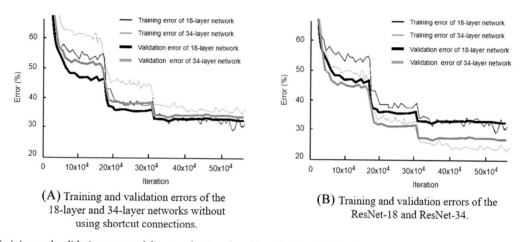

(A) Training and validation errors of the 18-layer and 34-layer networks without using shortcut connections.

(B) Training and validation errors of the ResNet-18 and ResNet-34.

FIG. 8.13   Training and validation errors of the neural networks with and without using shortcut connections on ImageNet [5].

After Microsoft proposed the ResNet architecture, many more improved Residual Blocks were introduced. The following is a summary of each:

- ResNeXt [11]: Published in 2017, introducing a block of ResNex with a new dimension called "cardinality" in addition to the depth and width dimensions for building an image classification network model
- SENet [12]: Published in 2017, introducing a "Squeeze-and-Excitation" (SE) block being able to perform dynamic channel-wise feature recalibration for improving the representational capacity of the neural network
- SKNet [13]: Published in 2019, proposing a dynamic selection mechanism for automatically choosing the size of the receptive field in the neural network

8.1.6 Comparison of network architectures

Fig. 8.14 presents the results of AlexNet, VGG, GoogleNet, and ResNet trained on the ImageNet dataset. As shown, ResNet, the deepest network architecture with 152 layers, achieved the least top-5 errors on the ImageNet test set, followed by GoogLeNet, VGG-19, and AlexNet.

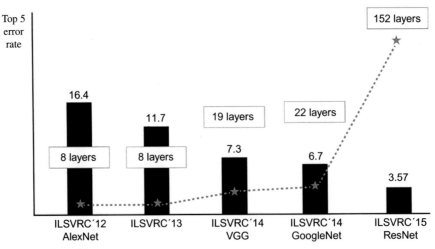

FIG. 8.14 Top-5 error CNNs on ImageNet test set.

8.2 Experiment: Implementation of inception-v3 network

This chapter presents how to implement the Inception-v3 network for image classification. The objective can be accomplished by using the TensorFlow library introduced in the previous chapters; however, it takes time because of complex network architecture. The following section introduces two methods for building the Inception-v3 network faster and more effectively: Keras Applications and TensorFlow Hub. Fig. 8.15 shows an example prediction of the Inception-v3 network.

- African elephant ： 80.37%
- Tusker ： 12.16%
- Indian elephant ： 0.42%

FIG. 8.15 The African elephant image prediction of the Inception-v3 network.

8.2.1 Keras applications

Keras Applications is the applications module of the Keras library, providing network declarations and pre-trained weights for popular architectures such as Xception, VGG, ResNet, Inception, and so on. Table 8.1 lists the accuracy of some networks on the ImageNet validation set used in the 2012 ILSVRC competitions obtained by employing Keras Applications.

TABLE 8.1 Network architectures provided by Keras applications.

Model	Top-1 accuracy	Top-5 accuracy	Parameters	Depth
Xception [9]	0.790	0.945	22,910,480	126
VGG16 [3]	0.713	0.901	138,357,544	23
VGG19 [3]	0.713	0.900	143,667,240	26
ResNet50 [5]	0.749	0.921	25,636,712	168
Inception-v3 [7]	0.779	0.937	23,851,784	159
InceptionResNetV2 [14]	0.803	0.953	55,873,736	572
MobileNet [15]	0.704	0.895	4,253,864	88
MobileNetV2 [16]	0.713	0.901	3,538,984	88
DenseNet121 [17]	0.750	0.923	8,062,504	121
DenseNet169 [17]	0.762	0.932	14,307,880	169
DenseNet201 [17]	0.773	0.936	20,242,984	201
NASNetMobile [18]	0.744	0.919	5,326,716	–
NASNetLarge [18]	0.825	0.960	88,949,818	–

- Top-1: if the prediction result of the model with the highest probability is the same as the answer, the prediction is correct
- Top-5: if the network predicts five results with the highest probability, one of which is the same as the answer, it is the correct prediction
- Parameters: the number parameters of the network
- Depth: the number of layers of the network, including convolutional layers, pooling layers, activation layers, and so on

The following program uses Keras Applications to build the Inception-v3 network as an example. Fig. 8.16 is a flowchart of the source code.

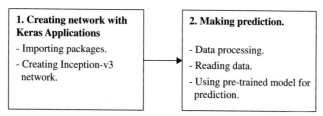

FIG. 8.16 Flowchart of the source code for building and testing the Inception-v3 network with Keras Applications.

1. Creating network with Keras Applications

    a) Import packages

```
import tensorflow as tf
import numpy as np
```

    b) Create Inception-v3 network

        ■ Create Inception-v3 network and load pre-trained weights:

```
model = tf.keras.applications.InceptionV3(include_top=True, weights='imagenet',
input_tensor=None, input_shape=None, pooling=None, classes=1000)
```

Arguments

- include_top:
  - True (default): including the fully connected layer at the top of the network
  - False: does not include the fully connected layer
- weights:
  - None: random initialization
  - Imagenet (default): using pre-trained weights on ImageNet

- input_tensor (optional): used for sharing inputs between multiple networks; "None" is set as the default
- input_shape (optional): input shape of (299, 299, 3); if "input_tensor" is used, "input_shape" is omitted
- pooling (optional): for feature extraction when "include_top" is set to False
  - None (default): the output of the last convolutional block is used as the output of the network
  - avg: using the global average pooling at the output of the last convolutional block
  - max: using global max pooling at the output of the last convolutional block
- Classes (optional): the number of classes; the default is 1000. If it is necessary to customize the number of classes, it is required to set "include_top" to True and "weights" to None

■ Display Inception-v3 architecture through "model.summary":

```
model.summary()
```

Result:

```
Model: "inception_v3"

Layer (type)                    Output Shape          Param #     Connected to
==================================================================================================
input_1 (InputLayer)            [(None, 299, 299, 3)  0

conv2d (Conv2D)                 (None, 149, 149, 32)  864         input_1[0][0]

batch_normalization (BatchNorma (None, 149, 149, 32)  96          conv2d[0][0]

activation (Activation)         (None, 149, 149, 32)  0           batch_normalization[0][0]

conv2d_1 (Conv2D)               (None, 147, 147, 32)  9216        activation[0][0]

batch_normalization_1 (BatchNor (None, 147, 147, 32)  96          conv2d_1[0][0]

activation_1 (Activation)       (None, 147, 147, 32)  0           batch_normalization_1[0][0]

conv2d_2 (Conv2D)               (None, 147, 147, 64)  18432       activation_1[0][0]
```

■ Write the data for displaying on TensorBoard:

```
model_tb = tf.keras.callbacks.TensorBoard(log_dir='lab8-logs-inceptionv3-keras')
model_tb.set_model(model)
```

■ Open TensorBoard through command line to view the Inception-v3 architecture:

```
# Open TensorBoard directly on the jupyter notebook
%load_ext tensorboard
%tensorboard --port 9600 --logdir lab8-logs-inceptionv3-keras
```

Result:

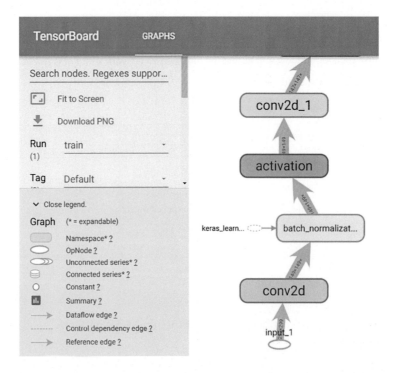

2. Making prediction
   a) Data processing
      For each network model, Keras Applications provides the corresponding functions for processing data.
      - preprocess_input: for processing input data following the format the network model requires. The input format of each model can be different. For example, some models require input image with the values range from −1 to +1, others use the value range from 0 to 1, and so on.
      - decode_predictions: for decoding the predictions of the pre-trained models and mapping the prediction values to the actual class names.

      Import data pre-processing and decoding functions

      ```
      from tensorflow.keras.applications.inception_v3 import preprocess_input
      from tensorflow.keras.applications.inception_v3 import decode_predictions
      ```

   b) Reading data

      - Create a function for reading data:

      ```
      def read_img(img_path, resize=(299,299)):
          #Read file
          img_string = tf.io.read_file(img_path)
      ```

      ```
          # Decode files in image format
          img_decode = tf.image.decode_image(img_string)
          # Resize the image to the network input size
          img_decode = tf.image.resize(img_decode, resize)
          # increase the dimension of the image to 4 (batch, height, width, channels)
          img_decode = tf.expand_dims(img_decode, axis=0)
          return img_decode
      ```

      - Read an image by using "read_img" funtion:

      ```
      img_path = 'image/elephant.jpg'
      # Read the image through the function just created
      img = read_img(img_path)
      # Display the image
      plt.imshow(tf.cast(img, tf.uint8)[0])
      ```

      Result:

      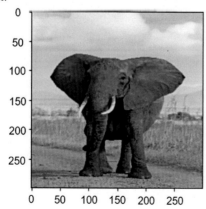

c) Using pre-trained model for prediction

```
img = preprocess_input(img)  # Image pre-processing
preds = model.predict(img)   # image prediction
print("Predicted:", decode_predictions(preds, top=3)[0])  # Show the three categories with
the highest output prediction
```

Result: Predicted: [('n02504458', 'African_elephant', 0.8037859), ('n01871265', 'tusker', 0.12163948), ('n02504013', 'Indian_elephant', 0.0042992835)]

8.2.2 TensorFlow Hub

TensorFlow Hub is a repository where the trained machine learning models are available for downloading and reusing with a minimum amount of code. There are various pre-trained models provided through the "tfhub.dev" repository such as image classification models and text embedding models. The official website of TensorFlow Hub is https://tfhub.dev/, as shown in Fig. 8.17.

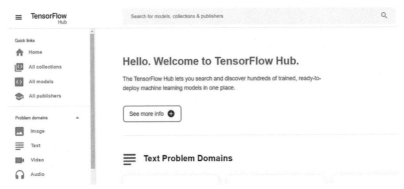

FIG. 8.17　TensorFlow Hub website.

To install TensorFlow Hub, please use the following command line:

```
pip install tensorflow-hub
```

The following program uses TensorFlow Hub to build the Inception-v3 network as an example. Fig. 8.18 is a flowchart of the source code.

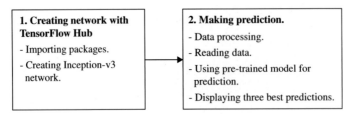

FIG. 8.18　Flowchart of the source code for building and testing the Inception-v3 network with TensorFlow Hub.

1. Creating network with TensorFlow Hub

   (a) Import packages

```
import tensorflow_hub as hub
```

   (b) Cerate Inception-v3 network
   - Step 1: Go to the website https://tfhub.dev/ and click the "classification" label as shown in Fig. 8.19.

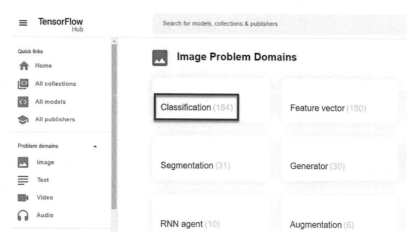

FIG. 8.19 Selection of classification label on TensorFlow Hub.

   - Step 2: Select Inception-v3, as shown in Fig. 8.20.

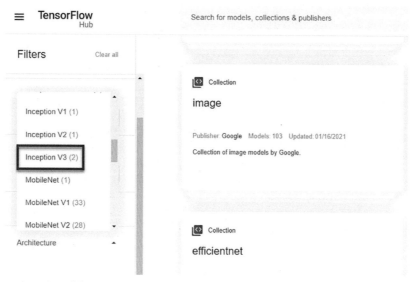

FIG. 8.20 Selection of Inception-v3 model in TensorFlow Hub.

- Step 3: Select the TF2 version of Inception-v3 network, as shown in Fig. 8.21.

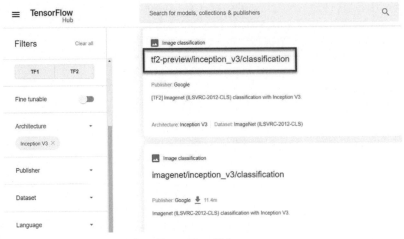

FIG. 8.21   Selection of version for Inception-v3 network on TensorFlow Hub.

- Step 4: Click "Copy URL" button to copy the URL for downloading the pre-trained Inception-v3 on TensorFlow Hub, as shown in Fig. 8.22.

FIG. 8.22   Information of Inception-v3 on TensorFlow Hub.

- Step 5: Create the Inception-v3 network through the URL copied in step 4

```
# URL of the pre-trained Inception-v3 model
module_url = " https://tfhub.dev/google/tf2-preview/inception_v3/classification/2"
# Buiding Inception-v3
model = tf.keras.Sequential([
    # hub.KerasLayer wraps the saved Inception-v3 model as a Keras layer
    hub.KerasLayer(module_url,
            input_shape=(299, 299, 3),
                output_shape=(1001, ),
                name='Inception_v3')
])
```

- Step 6: Display the architecture of Inception-v3 through model.summary

```
model.summary()
```

Result:

```
Model: "sequential"

_____
Layer (type)                Output Shape              Param #
================================================================
Inception_v3 (KerasLayer)   (None, 1001)              23853833
================================================================
Total params: 23,853,833
Trainable params: 0
Non-trainable params: 23,853,833
_____
```

2. Making prediction

(a) Reading data

- Create a function for reading data:

```
def read_img(img_path, resize=(299,299)):
    # Read file
    img_string = tf.io.read_file(img_path)
    # Decode files in image format
    img_decode = tf.image.decode_image(img_string)
    # Resize the image to the Inception-v3 input size
    img_decode = tf.image.resize(img_decode, resize)
    # Normalization
    img_decode = img_decode / 255.0
    # expand dimension
    img_decode = tf.expand_dims(img_decode, axis=0)
    return img_decode
```

- Reading the actual class names:

```
labels_path=tf.keras.utils.get_file('ImageNetLabels.txt',
    'https://storage.googleapis.com/download.tensorflow.org/data/ImageNetLabels.txt')
# Reading the data
with open(labels_path) as file:
    lines = file.read().splitlines()
# Displaying data
print(lines)
imagenet_labels = np.array(lines)
```

Result:

```
Downloading data from https://storage.googleapis.com/download.tensorflow.org/data/ImageNetLabels.txt
16384/10484 [==================================] - 0s 0us/step
['background', 'tench', 'goldfish', 'great white shark', 'tiger shark', 'hammerhead', 'electric ra
y', 'stingray', 'cock', 'hen', 'ostrich', 'brambling', 'goldfinch', 'house finch', 'junco', 'indigo
bunting', 'robin', 'bulbul', 'jay', 'magpie', 'chickadee', 'water ouzel', 'kite', 'bald eagle', 'vul
ture', 'great grey owl', 'European fire salamander', 'common newt', 'eft', 'spotted salamander', 'ax
olotl', 'bullfrog', 'tree frog', 'tailed frog', 'loggerhead', 'leatherback turtle', 'mud turtle', 't
errapin', 'box turtle', 'banded gecko', 'common iguana', 'American chameleon', 'whiptail', 'agama',
'frilled lizard', 'alligator lizard', 'Gila monster', 'green lizard', 'African chameleon', 'Komodo d
ragon', 'African crocodile', 'American alligator', 'triceratops', 'thunder snake', 'ringneck snake',
'hognose snake', 'green snake', 'king snake', 'garter snake', 'water snake', 'vine snake', 'night sn
ake', 'boa constrictor', 'rock python', 'Indian cobra', 'green mamba', 'sea snake', 'horned viper',
'diamondback', 'sidewinder', 'trilobite', 'harvestman', 'scorpion', 'black and gold garden spider',
'barn spider', 'garden spider', 'black widow', 'tarantula', 'wolf spider', 'tick', 'centipede', 'bla
ck grouse', 'ptarmigan', 'ruffed grouse', 'prairie chicken', 'peacock', 'quail', 'partridge', 'Afric
an grey', 'macaw', 'sulphur-crested cockatoo', 'lorikeet', 'coucal', 'bee eater', 'hornbill', 'hummi
ngbird', 'jacamar', 'toucan', 'drake', 'red-breasted merganser', 'goose', 'black swan', 'tusker', 'e
chidna', 'platypus', 'wallaby', 'koala', 'wombat', 'jellyfish', 'sea anemone', 'brain coral', 'flatw
orm', 'nematode', 'conch', 'snail', 'slug', 'sea slug', 'chiton', 'chambered nautilus', 'Dungeness c
```

- Read an image by using "read_img" function

```
img_path = 'image/elephant.jpg'
# Read image
img = read_img(img_path)
# Display image
plt.imshow(img[0])
```

Result:

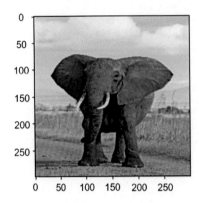

(b) Using pre-trained model for prediction

```
preds = model.predict(img)  # image prediction
index = np.argmax(preds)  # Get the Index with the largest prediction result
print("Predicted:", imagenet_labels[index])  # print the predicted result
```

Result: Predicted: African elephant

(c) Display three best predictions

```
# Get the three indexes with the largest prediction results
top3_indexs = np.argsort(preds)[0, ::-1][:3]
# mapping the prediction values to the actual class names.
print("Predicted:", imagenet_labels[top3_indexs])
```

Result: Predicted: ['African elephant' 'tusker' 'Indian elephant']

References

[1] Y. Lecun, L. Bottou, Y. Bengio, P. Haffner, Gradient-based learning applied to document recognition, Proc. IEEE 86 (11) (1998) 2278–2324.

[2] A. Krizhevsky, I. Sutskever, G.E. Hinton, Imagenet classification with deep convolutional neural networks, Adv. Neural Inf. Proces. Syst. (2012) 1097–1105.

[3] K. Simonyan, A. Zisserman, Very deep convolutional networks for large-scale image recognition, in: International Conference on Learning Representations, 2015, pp. 1–14.

[4] C. Szegedy, et al., Going deeper with convolutions, in: IEEE Conference on Computer Vision and Pattern Recognition, 2015, , pp. 1–9.

[5] K. He, X. Zhang, S. Ren, J. Sun, Deep residual learning for image recognition, Proc. IEEE Conf. Comput. Vis. Pattern Recognit. (2016) 770–778.

[6] S. Ioffe, C. Szegedy, Batch normalization: accelerating deep network training by reducing internal covariate shift, in: International Conference on Machine Learning, 2015, , pp. 448–456.

[7] C. Szegedy, V. Vanhoucke, S. Ioffe, J. Shlens, Z. Wojna, Rethinking the inception architecture for computer vision, in: Proceedings of the IEEE Conference on Computer Vision and Pattern Recognition, 2016.

[8] C. Szegedy, S. Ioffe, V. Vanhoucke, A. Alemi, Inception-V4, inception-ResNet and the impact of residual connections on learning, in: Association for the Advancement of Artificial Intelligence, 2017, , pp. 1–3.

[9] F. Chollet, Xception: deep learning with depthwise separable convolutions, in: Proceedings of the IEEE Conference on Computer Vision and Pattern Recognition, 2017, , pp. 1251–1258.

[10] K. He, J. Sun, Convolutional neural networks at constrained time cost, in: Proceedings of the IEEE Conference on Computer Vision and Pattern Recognition, 2015, , pp. 5353–5360.

[11] S. Xie, R. Girshick, P. Dollár, T. Zhuowen, K. He, Aggregated residual transformations for deep neural networks, in: Proceedings of the IEEE Conference on Computer Vision and Pattern Recognition, 2017, , pp. 1492–1500.

[12] J. Hu, L. Shen, G. Sun, Squeeze-and-excitation networks, in: Proceedings of the IEEE Conference on Computer Vision and Pattern Recognition, 2018, , pp. 7132–7141.

[13] X. Li, W. Wang, X. Hu, J. Yang, Selective kernel networks, in: Proceedings of the IEEE Conference on Computer Vision and Pattern Recognition, 2019, , pp. 510–519.

[14] C. Szegedy, S. Ioffe, V. Vanhoucke, A. Alemi, Inception-v4, inception-resnet and the impact of residual connections on learning. arXiv preprint arXiv:1602.07261, (2016).

[15] A.G. Howard, M. Zhu, B. Chen, D. Kalenichenko, W. Wang, T. Weyand, M. Andreetto, H. Adam, Mobilenets: efficient convolutional neural networks for mobile vision applications. arXiv preprint arXiv:1704.04861, (2017).

[16] M. Sandler, A. Howard, M. Zhu, A. Zhmoginov, L.-C. Chen, Mobilenetv2: inverted residuals and linear bottlenecks, in: Proceedings of the IEEE Conference on Computer Vision and Pattern Recognition, 2018, , pp. 4510–4520.

[17] G. Huang, Z. Liu, L. Van Der Maaten, K.Q. Weinberger, Densely connected convolutional networks, in: Proceedings of the IEEE Conference on Computer Vision and Pattern Recognition, 2017, , pp. 4700–4708.

[18] B. Zoph, V. Vasudevan, J. Shlens, V.L. Quoc, Learning transferable architectures for scalable image recognition, in: Proceedings of the IEEE Conference on Computer Vision and Pattern Recognition, 2018, , pp. 8697–8710.

9

Transfer learning

9.1 Transfer learning

9.1.1 Introduction to transfer learning

In Chapter 4, Section 4.1.3., we explained and demonstrated that the features extracted by convolutional layers of different depths of the network are different. In particular, the shallower layers are mainly responsible for extracting simple features of the input such as edges and lines, while the deeper layers are based on the extracted features from the previous layers, so they can recognize more specific features such as nose, eyes, ears, and so on, as shown in Fig. 9.1.

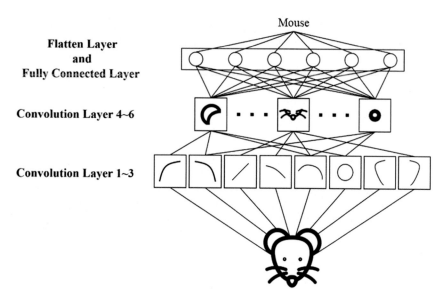

FIG. 9.1 Example of a convolutional neural network (CNN).

The learned feature of convolutional neural networks (CNNs) can be applied to different tasks. For example, by using the ImageNet dataset containing tens of thousands of images to train the CNNs, a large amount of training data allows the network models to learn very diverse features. When we have a new dataset and want to solve a task of interest, instead of training the entire network model from scratch, we use a trained model on a large dataset such as ImageNet, called a pre-trained model, as an initialization or fixed feature extractor for the new task. This training method is called transfer learning [1–9].

Transfer learning is actually very similar to our human learning method; the knowledge gained by completing a task can also be used to solve other related tasks. The more related the tasks are, the easier the knowledge is transferred. For example, when we know how to ride a bicycle, it will be easier for us to learn to ride a motorcycle. According to the experimental results of many public works [10–14], using the transfer learning method for training neural network models can help to obtain better performance while requiring less training data and training time compared with the method of training from scratch. Fig. 9.2 shows how training from scratch differs from transfer learning. As shown, in training from scratch, the models are trained separately for specific tasks and no knowledge is retained for transferring from one model to another. In transfer learning, learning of new tasks is based on the learned knowledge from previous tasks.

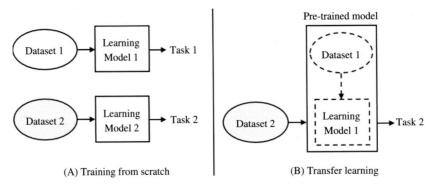

FIG. 9.2   Diagram of training a model from scratch and training a model using transfer learning.

9.1.2 Transfer learning strategies

There are four transfer learning strategies that can be applied based on the following two characteristics of the training data.

- The size of training data: If a new dataset contains tens of thousands of training samples, it is considered a large dataset. If the new dataset only consists of hundreds or thousands of training samples, it is considered a small dataset.
- The similarity of data: This is the similarity between the new training data and the data used for the pre-trained model. For example, cats and tigers are considered as similar samples, while cats and tables are considered as different samples.

(1) The first strategy: training network models with small dataset and similar training samples

Because a huge network model is trained on a small dataset, the overfitting problem is prone to occur. Thus, when applying transfer learning, the weights of the pre-trained model must remain unchanged during training process. Since the new dataset has similarity to the dataset used for the pre-trained model, they have similar features in hidden layers of the network, especially in the deeper layers. Therefore, the layers for feature extraction do not need to change. To learn a new task, it is only required to make changes to the last layer that handles output features of the network model. The new network model is built based on the architecture of the pre-trained model by removing the output layer (last layer), then adding new layers for the new task. The steps for transfer learning are as follows:

- Step 1 (removing the network layer): Choose to remove the last layer of the pre-trained model, as shown in Fig. 9.3.

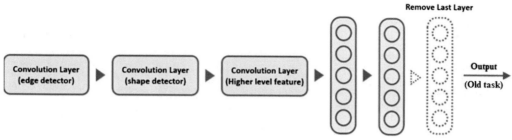

FIG. 9.3 Removing the last layer of the pre-trained model (1).

- Step 2 (adding new layer): Add one or multiple layers to the top of the original network structure for the new task, as shown in Fig. 9.4.

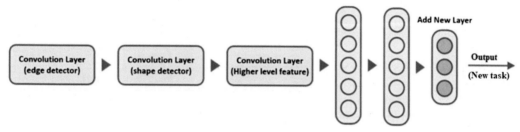

FIG. 9.4 Adding layers for the new task (1).

- Step 3 (training new model): During training the new network model, the weights of most network layers are fixed; only newly added layers are trained, as shown in Fig. 9.5.

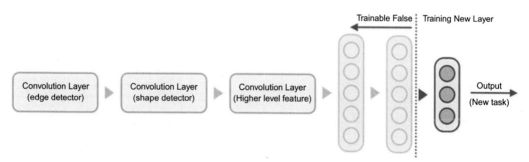

FIG. 9.5 Freezing layers for training (1).

(2) The second strategy: training network models with small datasets and dissimilar training samples

Because of using a small dataset for training a network model, the weights of the pre-trained model should be left unchanged to prevent the overfitting problem. Since the samples of the new dataset are different from these of the dataset used for the pre-trained model, they have similar features only in the shallower layers, while their features in the deeper layers of the network model are very different. To build a network model based on the architecture of the pre-trained model for learning a new task, only the shallower layers are retained, whereas the deeper layers should be removed, and the output layer is replaced with a new one. The steps for transfer learning are as follows:

- Step 1 (removing the network layer): Remove most of the deep layers of the pre-trained model and leave only some shallow layers that extract simple features such as lines and edges, as shown in Fig. 9.6.

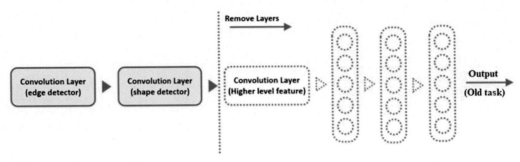

FIG. 9.6 Removing layers of the pre-trained model.

- Step 2 (adding new layer): Choose to add new layers to the top of the original network architecture for the new task, as shown in Fig. 9.7.

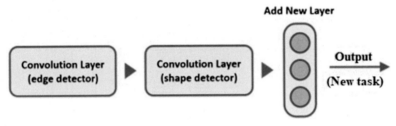

FIG. 9.7 Adding layers for the new task (2).

- Step 3 (training new model): During training the new network model, the weights of most network layers are fixed; only newly added layers are trained, as shown in Fig. 9.8.

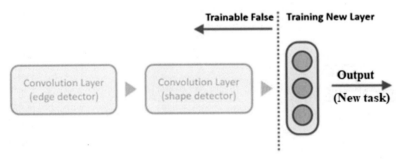

FIG. 9.8 Freezing layers for training (2).

(3) The third strategy: training network models with a large dataset and similar training samples

Using a large dataset allows us to train the entire pre-trained model or a new network model from scratch. The following provides two training methods:

(a) Using pre-trained model

Because the new dataset has similarity to the dataset used for the pre-trained model, they have similar features in hidden layers of the network, especially in the deeper layers. Therefore, the layers for feature extraction do not need to change. To learn a new task, it is only required to make changes to the last layer that handles the output of the network model. The new model is built based on the architecture of the pre-trained model by replacing the last layer with new one. Finally, because the dataset is large enough, the weights of the newly added layers as well as some or all other layers of the new network model can be optimized to obtain better performance. The steps for fine-tuning are as follows:

- Step 1 (removing the network layer): Choose to remove the last layer of the pre-trained model, as shown in Fig. 9.9.

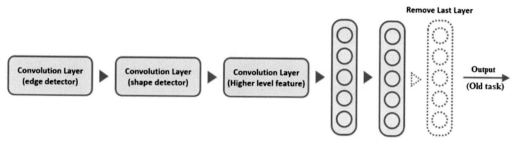

FIG. 9.9 Removing the last layer of the pre-trained model (2).

- Step 2 (adding new layers): Choose to add one or more layers to the top of the original network architecture for a new task, as shown in Fig. 9.10.

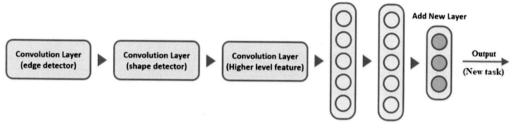

FIG. 9.10 Adding layer for the new task (3).

- Step 3 (training new model):
  - First method: The weights of most network layers are fixed; only newly added layers are trained, as shown in Fig. 9.11.

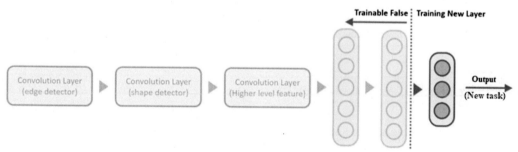

FIG. 9.11 Freezing layers for training (3).

- Second method: Train or fine-tune through the entire network model on the new dataset, as show in Fig. 9.12.

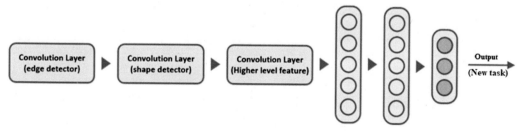

FIG. 9.12 Training or fine-tuning the entire network model (1).

(b) Creating a new network model

Because the dataset is large enough, it allows us to create a new network model and train this model from scratch, as shown in Fig. 9.13.

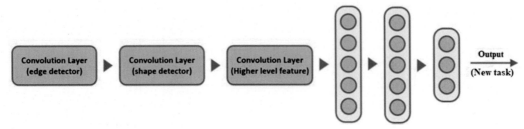

FIG. 9.13   Training the network model from scratch (1).

(4) The fourth strategy: training network models with large dataset and dissimilar training samples

Using a larger dataset for training a network model can avoid the overfitting problem. Therefore, the entire pre-trained network model or a new network model can be trained on the new dataset for improving performance. The following provides two training methods.

(a) Using a pre-trained model

Due to the difference between the new dataset and the dataset used for the pre-trained model, their features are not similar in most layers of the network model. However, in practice, it is still very often beneficial to initialize with weights from a pre-trained model for training a new network model. The steps for the training model are as follows:

- Step 1 (removing the network layer): Choose to remove the last layer of the pre-trained model, as shown in Fig. 9.14.

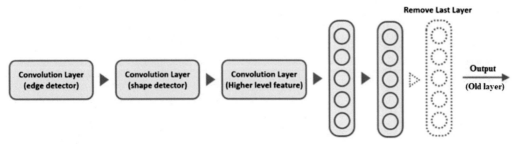

FIG. 9.14   Removing the last layer of the pre-trained model (3).

- Step 2 (adding new layer): Choose to add one or more layers to the top of the original network architecture for a new task, as shown in Fig. 9.15.

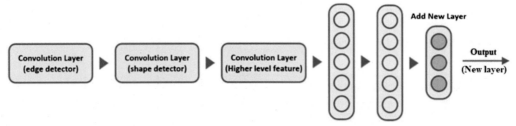

FIG. 9.15   Adding layers for the new task (4).

- Step 3 (training new model): Train or fine-tune the entire network model on the new larger dataset with weights initialization from the pre-trained model, as shown in Fig. 9.16.

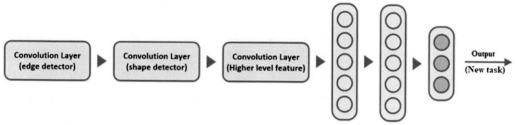

FIG. 9.16   Training or fine-tuning the entire network model (2).

(b) Creating a new network model

Because the dataset is large enough, it allows us to create a new network model and train this model from scratch, as shown in Fig. 9.17.

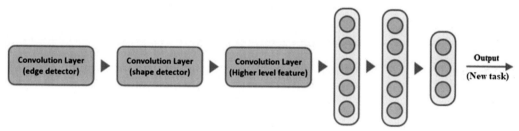

FIG. 9.17   Training the network model from scratch (2).

9.2 Experiment: Using Inception-v3 for transfer learning

This section introduces how to use the Inception-v3 network for improving image classification performance on the Dogs vs. Cats dataset. We conduct two training scenarios. In the first scenario, we build the Inception-v3 model, named Model-1, through Keras Applications and train it from scratch with random weight initialization. In the second scenario, we build a new network model, named Model-2, by using the Inception-v3 network loaded through TensorFlow Hub for feature extraction and adopting two fully connected layers for classification. By applying transfer learning to Model-2, the weights of most of the layers are fixed; only two newly added fully connected layers are trained on the Dogs vs. Cats dataset to learn a new task. The experimental results show that Model-2 obtains greater accuracy than Model-1 does.

9.2.1 Introduction to Dogs vs. Cats dataset

The Dogs vs. Cats dataset, which includes two classes of dog and cat, was introduced for a Kaggle machine learning competition in 2013. The Kaggle provides 25,000 labeled images of dogs and cats for training and 12,500 unlabeled images for testing. Fig. 9.18 shows some of the images from the Dogs vs. Cats dataset. To build an optimized input pipeline in the experiment, the Dogs vs. Cats dataset is loaded from the TensorFlow dataset through the "tf.data. Datasets" application programming interface (API) for training and testing network models.

FIG. 9.18   Images from Dogs vs. Cats dataset.

9.2.2 Code examples

Fig. 9.19 is a flowchart of the source code for images classification models.

1. **Preparing data**	2. **Building and training network models**	3. **Display results**
- Importing packages. - Loading data. - Setting data (training data, validation data, and test data).	- Build Model-1. - Build Model-2. - Create Callback function. - Set optimizer, loss function, metric function. - Training Model-1, Model-2.	- The accuracy of Model-1 and Model-2 on test data. - Comparison results of two models.

FIG. 9.19   Flowchart of the source code for images classification models.

1. Preparing dataset

a) Import packages

```
import os
import numpy as np
import tensorflow as tf
import tensorflow_hub as hub
import tensorflow_datasets as tfds
from tensorflow import keras
from tensorflow.keras import layers
import matplotlib.pyplot as plt
from preprocessing import flip, color, rotate, zoom
```

b) Loading data

- Load Cats_vs_Dog dataset:

```
# Divide data with ratio of 8:1:1 for training, validation and testing
train_split, valid_split, test_split = tfds.Split.TRAIN.subsplit([80, 10, 10])
# load the training set
train_data, info = tfds.load("cats_vs_dogs", split=train_split, with_info=True)
# load validation set
valid_data = tfds.load("cats_vs_dogs", split=valid_split)
# load test set
test_data = tfds.load("cats_vs_dogs", split=test_split)
```

- View name of class and create decoder:

```
print(info.features['label'].names)  # display name of class
decoder = info.features['label'].names  # create decoder
```
Result: ['cat', 'dog']

- Display the image from dataset:

```
for data in train_data.take(1):
    img = data['image']  # read image
    label = data['label']  # read lable
# get the class
plt.title(decoder[label])
# Display image
plt.imshow(img)
```
Result:

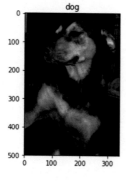

c) Setting data

- Creating function for data augmentation:

```
input_shape = (299, 299)  # set the input size
def parse_aug_fn(dataset):
    "
    Image Augmentation function
    "
    # Image standardization
    x = tf.cast(dataset['image'], tf.float32) / 255.
    x = tf.image.resize(x, input_shape)
    # Random horizontal flip
    x = flip(x)
    # color conversion (50%)
    x = tf.cond(tf.random.uniform([], 0, 1) > 0.5, lambda: color(x), lambda: x)
    # Image rotation (25%)
    x = tf.cond(tf.random.uniform([], 0, 1) > 0.75, lambda: rotate(x), lambda: x)
    # image scaling (50%)
    x = tf.cond(tf.random.uniform([], 0, 1) > 0.5, lambda: zoom(x), lambda: x)
    return x, dataset['label']


def parse_fn(dataset):
    # Image standardization
    x = tf.cast(dataset['image'], tf.float32) / 255.
    x = tf.image.resize(x, input_shape)
    return x, dataset['label']
```

- Setting data for training, validation, and testing:

```
AUTOTUNE = tf.data.experimental.AUTOTUNE  # Automatic adjustment mode
buffer_size = 1000  # Because the image is larger, the cache space is set to 1000.
batch_size = 64  # Batch size


# Training data
train_data = train_data.map(map_func=parse_aug_fn, num_parallel_calls=AUTOTUNE)
# shuffle training data
train_data = train_data.shuffle(buffer_size)
# Set batch size and turn on prefetch mode
train_data = train_data.batch(batch_size).prefetch(buffer_size=AUTOTUNE)

#Validation data
valid_data = valid_data.map(map_func=parse_fn, num_parallel_calls=AUTOTUNE)
# Set batch size and turn on prefetch mode
valid_data = valid_data.batch(batch_size).prefetch(buffer_size=AUTOTUNE)

# Test data
test_data = test_data.map(map_func=parse_fn, num_parallel_calls=AUTOTUNE)
# Set batch size and turn on prefetch mode
test_data = test_data.batch(batch_size).prefetch(buffer_size=AUTOTUNE)
```

2. Building and training network models

a) Model-1: Training from scratch

- Create a storage directory for saving model:

```
model_dir = 'lab9-logs/models'  # set storage directory path
os.makedirs(model_dir)  # Create a storage directory
```

- Set callback function:

```
# Save training log
log_dir = os.path.join('lab9-logs', 'model-1')
model_cbk = keras.callbacks.TensorBoard(log_dir=log_dir)
# early stopping during training
model_esp = keras.callbacks.EarlyStopping(monitor='val_binary_accuracy',
                    patience=30,
                    mode='max')
```

- Create Inception-v3 network:

```
# Loading Inception-v3 network from keras.applications
base_model = tf.keras.applications.InceptionV3(include_top=False, # does not include the
fully connected layer.
                    weights=None, # random initialization
                    pooling='avg',
                    input_shape=input_shape+(3,))
# adding two fully connected layers to top of the base Inception-v3 model.
# using the Sigmoid activation function in the last layer
model_1 = tf.keras.Sequential([
    base_model,
    layers.Dense(128, activation='relu'),
    layers.Dense(1, activation='sigmoid')
])
```

- View model information via "model.summary" API:

```
model_1.summary()
```

Result:

```
Model: "sequential"
```

Layer (type)	Output Shape	Param #
inception_v3 (Model)	(None, 2048)	21802784
dense (Dense)	(None, 128)	262272
dense_1 (Dense)	(None, 1)	129

```
Total params: 22,065,185
Trainable params: 22,030,753
Non-trainable params: 34,432
```

- Set the optimizer, loss function, and metric function:

```
model_1.compile(keras.optimizers.Adam(),
        loss=keras.losses.BinaryCrossentropy(),
        metrics=[keras.metrics.BinaryAccuracy()])
```

- Training network model:

```
history = model_1.fit(train_data,
        epochs=200,
        validation_data=valid_data,
        callbacks=[model_cbk, model_esp])
```

Result:

```
Epoch 86/200
582/582 [==============================] - 290s 499ms/step - loss: 0.0359 - binary_accuracy: 0.9864 - val_loss: 0.0499 -
val_binary_accuracy: 0.9793
Epoch 87/200
582/582 [==============================] - 291s 499ms/step - loss: 0.0326 - binary_accuracy: 0.9880 - val_loss: 0.0673 -
val_binary_accuracy: 0.9750
Epoch 88/200
582/582 [==============================] - 291s 499ms/step - loss: 0.0386 - binary_accuracy: 0.9858 - val_loss: 0.0633 -
val_binary_accuracy: 0.9767
Epoch 89/200
582/582 [==============================] - 290s 498ms/step - loss: 0.0346 - binary_accuracy: 0.9865 - val_loss: 0.0738 -
val_binary_accuracy: 0.9707
Epoch 90/200
582/582 [==============================] - 290s 498ms/step - loss: 0.0361 - binary_accuracy: 0.9858 - val_loss: 0.0678 -
val_binary_accuracy: 0.9759
Epoch 91/200
582/582 [==============================] - 290s 498ms/step - loss: 0.0347 - binary_accuracy: 0.9868 - val_loss: 0.0744 -
val_binary_accuracy: 0.9728
```

b) Model-2: Transfer Learning

- Set callback function:

```
# Save training log
log_dir = os.path.join('lab9-logs', 'model-2')
model_cbk = keras.callbacks.TensorBoard(log_dir=log_dir)
# early stopping during training
model_esp = keras.callbacks.EarlyStopping(monitor='val_binary_accuracy',
                patience=30,
                    mode='max')
```

- Create new network model based on Inception-v3 architecture:

```
# pre-trained Inception-v3 model URL (the model does not contain a fully connected layer)
module_url = "https://tfhub.dev/google/tf2-preview/inception_v3/feature_vector/4"
# Create a new network model based on InceptionV3
model_2 = tf.keras.Sequential([
    hub.KerasLayer(module_url,
                    input_shape=(299, 299, 3),
                    output_shape=(2048,),
                    trainable=False),    # freeze weights
# add two fully connected layers, and use the Sigmoid activation function at the last layer.
    layers.Dense(128, activation='relu'),
    layers.Dense(1, activation='sigmoid')
])
```

- View model information via "model.summary":

```
model_2.summary()
```

Result:

```
Model: "sequential_1"

Layer (type)              Output Shape           Param #
=================================================================
keras_layer (KerasLayer)  (None, 2048)           21802784

dense_2 (Dense)           (None, 128)            262272

dense_3 (Dense)           (None, 1)              129
=================================================================
Total params: 22,065,185
Trainable params: 262,401
Non-trainable params: 21,802,784
```

- Set the optimizer, loss function, and metric function:

```
model_2.compile(keras.optimizers.Adam(),
            loss=keras.losses.BinaryCrossentropy(),
            metrics=[keras.metrics.BinaryAccuracy()])
```

- Training network model:

```
history = model_2.fit(train_data,
            epochs=200,
            validation_data=valid_data,
            callbacks=[model_cbk, model_esp])
```

Reuslt:

```
Epoch 20/200
582/582 [==============================] - 90s 155ms/step - loss: 0.0339 - binary_accuracy: 0.9864 - val_loss: 0.0180 - v
al_binary_accuracy: 0.9927
Epoch 21/200
582/582 [==============================] - 89s 154ms/step - loss: 0.0341 - binary_accuracy: 0.9864 - val_loss: 0.0184 - v
al_binary_accuracy: 0.9922
Epoch 22/200
582/582 [==============================] - 90s 154ms/step - loss: 0.0330 - binary_accuracy: 0.9867 - val_loss: 0.0204 - v
al_binary_accuracy: 0.9922
Epoch 23/200
582/582 [==============================] - 90s 155ms/step - loss: 0.0333 - binary_accuracy: 0.9879 - val_loss: 0.0205 - v
al_binary_accuracy: 0.9901
Epoch 24/200
582/582 [==============================] - 89s 154ms/step - loss: 0.0330 - binary_accuracy: 0.9868 - val_loss: 0.0192 - v
al_binary_accuracy: 0.9914
Epoch 25/200
582/582 [==============================] - 90s 154ms/step - loss: 0.0327 - binary_accuracy: 0.9869 - val_loss: 0.0236 - v
al_binary_accuracy: 0.9905
```

Supplementary explanation

In Chapter 8, the Inception-v3 network was loaded through TensorFlow Hub at https://tfhub.dev/google/tf2-preview/inception_v3/classification/4, and in this chapter, it is loaded at https://tfhub.dev/google/tf2-preview/inception_v3/feature_vector/4. The difference between the two loaded networks is that the Inception-v3 loaded in Chapter 8 contains the last classification layer (1000 categories), whereas this classification layer is removed in the latter network.

3. Comparison of Model-1 and Model-2

Both the best-trained weights of Model-1 and Model-2 are used for evaluating the Cats_vs_Dogs test set.

```
# Load the best trained weights of Model-1
model_1.load_weights(model_dir + '/Best-model-1.h5')
# Load the best trained weights of Model-2
model_2.load_weights(model_dir + '/Best-model-2.h5')
# Calculate the loss value and accuracy of Model-1 and Model-2
loss_1, acc_1 = model_1.evaluate(test_data)
loss_2, acc_2 = model_2.evaluate(test_data)
print("Model_1 Prediction: {}%".format(acc_1 * 100))
print("Model_2 Prediction: {}%".format(acc_2 * 100))
```

Result： Model_1 Prediction: 97.97413945198059%

Model_2 Prediction: 99.39655065536499%

The results show that Model-2 with transfer learning obtained a classification accuracy of 99.39%, which is greater than the 1.42% accuracy of Model-1 with training from scratch.

References

[1] J. Yosinski, J. Clune, Y. Bengio, H. Lipson, How transferable are features in deep neural networks? Adv. Neural Inf. Proces. Syst. (2014) 3320–3328.

[2] M. Oquab, L. Bottou, I. Laptev, J. Sivic, Learning and transferring mid-level image representations using convolutional neural networks, in: Proceedings of the IEEE Conference on Computer Vision and Pattern Recognition, Columbus, 2014, pp. 1717–1724.

[3] R. Mormont, P. Geurts, R. Marée, Comparison of deep transfer learning strategies for digital pathology, in: Proceedings of the IEEE Conference on Computer Vision and Pattern Recognition Workshops, 2018, pp. 2343–234309.

[4] S. Kornblith, J. Shlens, Q.V. Le, Do better ImageNet models transfer better? Proc. IEEE Conf. Comput. Vis. Pattern Recognit. (2019) 2661–2671.

[5] H. Shin, et al., Deep convolutional neural networks for computer-aided detection: CNN architectures, dataset characteristics and transfer learning, IEEE Trans. Med. Imaging 35 (5) (2016) 1285–1298.

[6] N. Tajbakhsh, et al., Convolutional neural networks for medical image analysis: full training or fine tuning? IEEE Trans. Med. Imaging 35 (5) (2016) 1299–1312.

[7] Y. Ganin, E. Ustinova, H. Ajakan, P. Germain, H. Larochelle, F. Laviolette, M. Marchand, V. Lempitsky, Domain-adversarial training of neural networks, J. Mach. Learn. Res. 17 (1) (2016) 2030–2096.

[8] D. Hendrycks, K. Lee, M. Mazeika, Using pre-training can improve model robustness and uncertainty, in: International Conference on Machine Learning, 2019.

[9] Z. Ding, Y. Fu, Deep transfer low-rank coding for cross-domain learning, IEEE Trans. Neural Netw. Learn. Syst. 30 (6) (2019) 1768–1779.

[10] F. Zhuang, et al., A comprehensive survey on transfer learning, Proc. IEEE 109 (1) (2021) 43–76.

[11] Z. Li, D. Hoiem, Learning without forgetting, IEEE Trans. Pattern Anal. Mach. Intell. 40 (12) (2018) 2935–2947.

[12] S.-C. Huang, T.-H. Le, D.-W. Jaw, DSNet: joint semantic learning for object detection in inclement weather conditions. IEEE Trans. Pattern Anal. Mach. Intell. (2021), https://doi.org/10.1109/TPAMI.2020.2977911.

[13] T.-H. Le, S.-C. Huang, D. Jaw, Cross-resolution feature fusion for fast hand detection in intelligent homecare systems, IEEE Sens. J. 19 (12) (2019) 4696–4704.

[14] Q.-V. Hoang, T.-H. Le, S.-C. Huang, An improvement of RetinaNet for hand detection in intelligent homecare systems, in: 2020 IEEE International Conference on Consumer Electronics - Taiwan (ICCE-Taiwan), 2020, pp. 1–2.

10

Variational auto-encoder

10.1 Introduction to auto-encoder

The aim of Auto-Encoder (AE) is to learn to compress data while minimizing errors in reconstructing them. To accomplish this, an AE is composed of two main parts: an encoder and a decoder [1–3]. The encoder is responsible for compressing the input into a lower-dimensional latent space representation. The latent representation is referred to as code, and the decoder is used to decode this code back to the input. In training the AE model, it is hoped that the output of the decoder and the input of the encoder are as similar as possible. For example, an MNIST handwritten image of size 28×28 is compressed into a 2D vector code by the encoder, and then the resulting code is decoded back to the original 28×28 image through the decoder. Fig. 10.1 presents the training schematic diagram of the AE model.

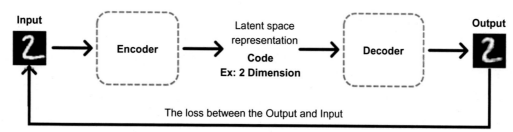

FIG. 10.1   Auto-Encoder (AE) training schematic diagram.

However, when training the AE model, there are no constraints on the latent representations generated by the encoder, so it cannot be ensured that every output generated by the decoder has meaningful attributes of the original data input. For example, 225 sets of 2D vector codes are produced by the encoder of an AE model, in which the values of codes are linearly sampled from -1.5 to $+1.5$. These 225 sets are sent to the decoder to generate 225 images. As observed in Fig. 10.2, we cannot recognize all characters in the output images except for images "0," "3," "5," and "8" in the lower right corner. In the next section, we introduce a Variational Auto-Encoder (VAE), which adds constraints on the latent representations to address this problem.

Principles and Labs for Deep Learning
https://doi.org/10.1016/B978-0-32-390198-7.00010-0

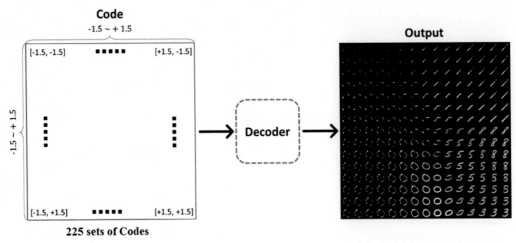

FIG. 10.2 Example of generated images of the Auto-Encoder (AE) model. Not all generated images can be recognized.

10.2 Introduction to variational auto-encoder

10.2.1 Introduction to VAE

VAE [4] is an advanced version of AE that learns a data-generating distribution, allowing it to take random codes or latent representations from latent distribution to generate output data that have similar characteristics to those of input data. Fig. 10.3 shows the differences between VAE and AE.

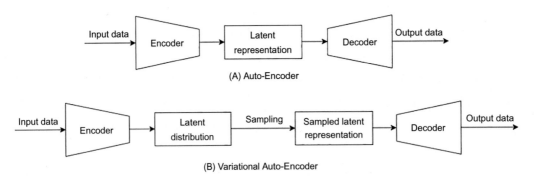

FIG. 10.3 Differences between Auto-Encoder (AE) and Variational Auto-Encoder (VAE).

Fig. 10.4 shows the operation concept of the VAE. As shown, the encoder receives input data and outputs two vectors including a vector of the mean (μ) and a vector of variance (σ^2) for generating latent distribution. Then, a code or latent representation is randomly sampled from the latent distribution to be passed through the decoder for generating output data.

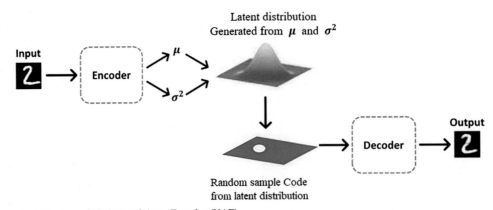

FIG. 10.4 Conceptual diagram of Variational Auto-Encoder (VAE).

As for why VAE has codes to be randomly sampled from the latent distributions, here is an intuitive example to explain. As shown in Fig. 10.5, two images A and B are used as the inputs of the encoder. At the output of the encoder, two codes, namely Code A and Code B are sampled and sent to the decoder to produce Output A and Output B, respectively. Because there is an intersection between two distributions, an additional Output C can be produced between Output A and Output B from this intersection. This design makes the outputs of the VAE have a continuous relationship, such as the continuous relationship between Output A, Output B, and Output C.

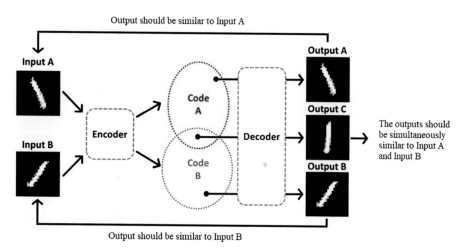

FIG. 10.5   Advantages of using latent distributions in VAE.

10.2.2 Operation of VAE

As introduced previously, instead of directly outputting the values for the latent space representation like AE, the encoder of a VAE outputs two vectors of the mean (μ) and variance (σ^2) for generating latent distributions. The decoder takes random sampled code from the latent distributions for reconstruction of the original input. When training the VAE model, the relationship of parameters in the model with respect to the final loss is calculated by the backpropagation algorithm. However, it is impossible to do this for the random sampling process because there is no value for computation. To overcome this problem, the VAE employed a reparametrization trick for sampling, in which an ε from a standard normal distribution is randomly sampled to combine with parameters μ and σ^2 of latent distribution for computing the output code of the encoder, which is defined as $C = exp(\sigma^2) * \varepsilon + \mu$, as shown in Fig. 10.6. Following this trick, μ and σ of the latent distribution can be optimized during the training process while

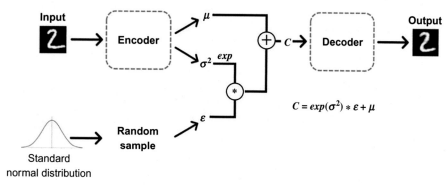

FIG. 10.6   Diagram of Variational Auto-Encoder (VAE).

still allowing to randomly sample from that distribution. Another more intuitive explanation for using the code $C = exp(\sigma^2) * \varepsilon + \mu$ is that μ is considered the code of the AE, this code is added a noise $(exp(\sigma^2) * \varepsilon)$ in the VAE to produce the code C and hope that C can still be decoded to the original input.

Taking a similar example as in Section 10.1, the encoder of the VAE produces 225 sets of 2D vector codes, and then these sets are passed through the decoder to generate 225 handwritten digit images, as shown in Fig. 10.7. As shown, each image is a digit that can be easy to recognize, and each is smoothly transformed into another.

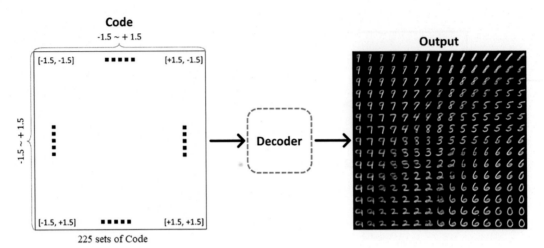

FIG. 10.7　Example of generated images of the Variational Auto-Encoder (VAE).

10.2.3 Variational auto-encoder loss function

The training goal of the VAE model is to ensure the predicted output and the input are as similar as possible. If the data contains image samples, binary cross-entropy (BCE) loss between each pixel of the reconstructed image and input image can be employed for training the model. This loss is also called reconstruction loss, which is described as follows:

$$Loss_{reconstruction} = \frac{1}{N}\sum_{i=1}^{N}\sum_{x=1}^{W}\sum_{y=1}^{H}\sum_{c=1}^{C} binary\_crossentropy\left(x_{x,y,c}, \hat{y}_{x,y,c}\right)$$

x: input image.
$\hat{y}$: output image or reconstructed image.
W: width of image.
H: height of image.
C: the number of image channels.
N: the amount of data in a batch.

However, if only reconstruction loss is used for training the VAE model, it is not enough. As shown in Fig. 10.6, $exp(\sigma^2)$ controls the scale of noise $(exp(\sigma^2) * \varepsilon)$, in which σ^2 is learned by the encoder. If the encoder learns to output $exp(\sigma^2)$ as 0, noise is not added to compute the output code. If there is no noise in computing the code, the output code in the VAE model is similar to that of the AE model, which means constraints on the encoded representations are not added to the VAE model. The graph of $exp(\sigma^2)$ is shown in Fig. 10.8, where, the smaller the value of σ^2 is, the closer to 0 the value of $exp(\sigma^2)$ becomes.

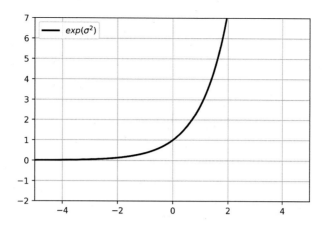

FIG. 10.8 The graph of $exp(\sigma^2)$.

In order to solve this problem, $Loss_{\sigma^2}$ is employed to limit the value of σ^2. If the value of $Loss_{\sigma^2}$ is equal to 0, σ^2 must be 0. At this time, the value of $exp(\sigma^2)$ is equal to 1, so the problem that the encoder updates in the direction of $exp(\sigma^2) = 0$ is solved. Fig. 10.9 shows the graph of $Loss_{\sigma^2}$; its formula is as follows:

$$Loss_{\sigma^2} = \frac{1}{2N} \sum_{i=1}^{N} \left[\exp\left(\sigma_i^2\right) - \left(1 + \sigma_i^2\right) \right]$$

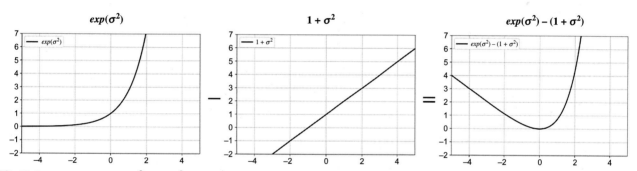

FIG. 10.9 The graph of $\exp(\sigma_i^2) - (1 + \sigma_i^2)$.

In addition, L2 regularization is applied for μ, which is formulated as follows:

$$Loss_u = \frac{1}{2N} \sum_{i=1}^{N} \mu_i^2$$

Finally, $Loss_{u,\sigma^2}$ is established by incorporating $Loss_u$ and $Loss_{\sigma^2}$, which is also known as Kullback–Leibler divergence loss (KL Loss) and is expressed as follows:

$$Loss_{u,\sigma^2} = \frac{1}{2N} \sum_{i=1}^{N} \left[\exp\left(\sigma_i^2\right) - \left(1 + \sigma_i^2\right) + \mu_i^2 \right]$$

μ: mean value (one of the outputs of encoder).
σ^2: variance (the other output of encoder).
N: batch size.

10.3 Experiment: Implementation of variational auto-encoder model

This section introduces an example program of building a VAE model, which is trained and tested on the MNIST handwritten digit dataset [5] for compressing and reconstructing images. Fig. 10.10 shows some generated images of the VAE model.

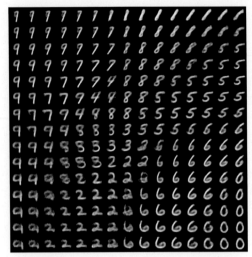

FIG. 10.10   Handwritten digital images generated by the Variational Auto-Encoder (VAE).

10.3.1 Create project

Because the VAE models in this chapter are much more complicated than the example programs in the previous chapters, we employ Pycharm IDE as a compiler to write the source codes and train the model. In the following, we outline the process of creating the project.

1. Create a new project: Click "File" → "New Project," as shown in Fig. 10.11.

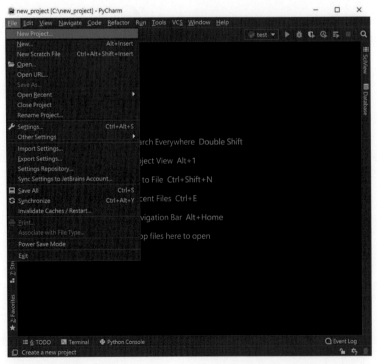

FIG. 10.11   Creating a new project on Pycharm.

2. Set a directory of the new project, as shown in Fig. 10.12.

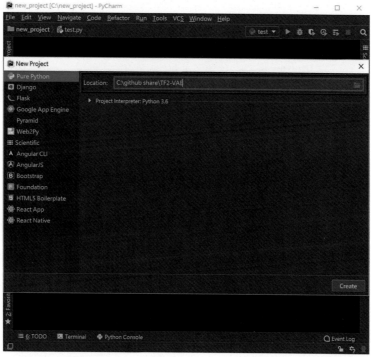

FIG. 10.12 Setting a new project directory.

3. Configure a Python interpreter: Open "Project Interpreter: Python 3.6," select the "Existing interpreter," and set Python Interpreter, as shown in Fig. 10.13.

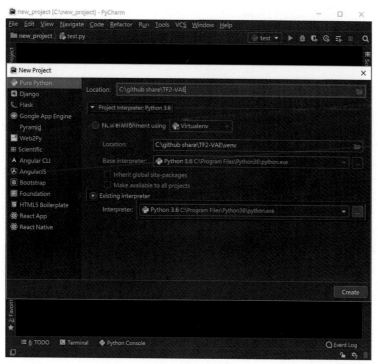

FIG. 10.13 Interpreter setting.

4. Create a project: Click "Create" button to create a new project, as shown in Fig. 10.14.

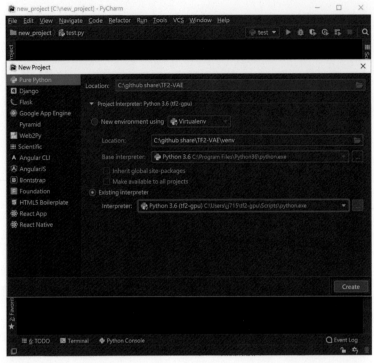

FIG. 10.14   Creating a project.

Supplementary explanation

The code examples of the VAE project in this chapter can be download at: https://github.com/taipeitechmmslab/MMSLAB-DL/tree/master, as shown in Fig. 10.15.

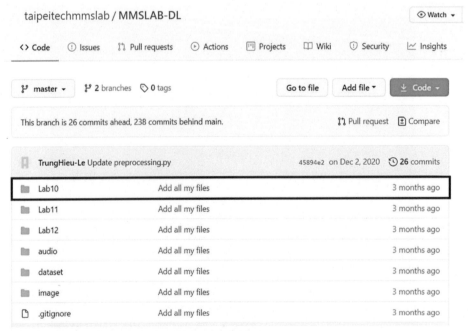

FIG. 10.15   The source code of Variational Auto-Encoder (VAE) on GitHub.

10.3.2 Introduction to the dataset

We use the MNIST handwritten digit dataset [5] for training and testing the VAE model in this chapter. The MNIST contains 60,000 training samples and 10,000 test samples, which are gray images with a size of 28×28. The dataset can be loaded through TensorFlow datasets as follows.

```python
import tensorflow_datasets as tfds

#Load training data
train_data, info = tfds.load("mnist", split= tfds.Split.TRAIN, with_info=True)
# Loading test data
test_data = tfds.load("mnist", split= tfds.Split.TRAIN)
# Display information of dataset
print(info)
```

Result:

```
tfds.core.DatasetInfo(
    name='mnist',
    version=1.0.0,
    description='The MNIST database of handwritten digits.',
    urls=['https://storage.googleapis.com/cvdf-datasets/mnist/'],
    features=FeaturesDict({
        'image': Image(shape=(28, 28, 1), dtype=tf.uint8),
        'label': ClassLabel(shape=(), dtype=tf.int64, num_classes=10),
    }),
    total_num_examples=70000,
    splits={
        'test': 10000,
        'train': 60000,
    },
    supervised_keys=('image', 'label'),
    citation="""@article{lecun2010mnist,
      title={MNIST handwritten digit database},
      author={LeCun, Yann and Cortes, Corinna and Burges, CJ},
      journal={ATT Labs [Online]. Available: http://yann. lecun. com/exdb/mnist},
      volume={2},
      year={2010}
    }""",
    redistribution_info=,
)
```

Fig. 10.16 shows some example images of the MNIST dataset.

FIG. 10.16 Handwritten digits from the MNIST dataset.

10.3.3 Building a Variational auto-encoder model

1. Directory and files

    Create a folder for storing python files of the VAE project, as shown in Fig. 10.17. Here, we present a brief intro-
    duction to the files.

    - train.py: file source code for training the VAE model
    - test.py: file source code for evaluating the VAE model
    - utils:
        - model.py: file source code for the VAE model and custom network layer
        - losses.py: file source code for custom loss function
        - callbacks.py: file source code for custom callback function

FIG. 10.17   Variational Auto-Encoder (VAE) project.

2. Implement the VAE model

    Fig. 10.18 is a flowchart of the source code for building the VAE model.

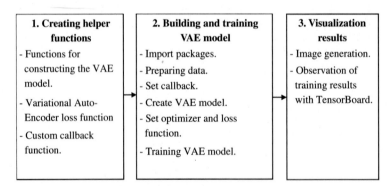

FIG. 10.18   Flowchart of the source code for the Variational Auto-Encoder (VAE) model.

(a) Creating helper functions
  - Functions for constructing the VAE model

The source code of the VAE model is written in the "models.py" file. Fig. 10.19 shows the architecture of the VAE.

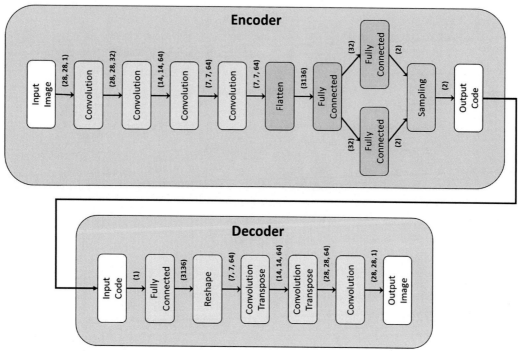

FIG. 10.19 The architecture of Variational Auto-Encoder (VAE).

Creating "create_vae_model" function.

```python
def create_vae_model(input_shape, latent_dim):
    # Define Encoder model
    img_inputs = keras.Input(input_shape)
    x = keras.layers.Conv2D(32, 3, padding='same', activation='relu')(img_inputs)
    x = keras.layers.Conv2D(64, 3, strides=2, padding='same', activation='relu')(x)
    x = keras.layers.Conv2D(64, 3, strides=2, padding='same', activation='relu')(x)
    x = keras.layers.Conv2D(64, 3, padding='same', activation='relu')(x)
    # The size of the feature map is saved before flattening,
    shape_before_flatten = x.shape
    x = keras.layers.Flatten()(x)
    x = keras.layers.Dense(32, 'relu')(x)
    # Output mean μ
    z_mean = keras.layers.Dense(latent_dim)(x)
    # Output variance σ²
    z_log_var = keras.layers.Dense(latent_dim)(x)
    # Custom network layer, explained below
    z = Sampling()([z_mean, z_log_var])
    # Create Encoder model
    encoder = keras.Model(inputs=img_inputs, outputs=z, name='encoder')
    encoder.summary()


    # For Decoder model
    latent_inputs = keras.Input((latent_dim,))
    x = keras.layers.Dense(np.prod(shape_before_flatten[1:]),
            activation='relu')(latent_inputs)
    # Reshape feature map
    x = keras.layers.Reshape(target_shape=shape_before_flatten[1:])(x)
    x = keras.layers.Conv2DTranspose(64, 3, 2, padding='same', activation='relu')(x)
    x = keras.layers.Conv2DTranspose(64, 3, 2, padding='same', activation='relu')(x)
    img_outputs = keras.layers.Conv2D(1, 3, padding='same', activation='sigmoid')(x)
    # Create Decoder model
    decoder = keras.Model(inputs=latent_inputs, outputs=img_outputs, name='decoder')
    decoder.summary()


    # Connect Encoder and Decoder to create VAE network model
    z = encoder(img_inputs)
    img_outputs = decoder(z)
    # Create VAE Model
    vae = keras.Model(inputs=img_inputs, outputs=img_outputs, name='vae')

    # Loss_{u,σ²} = (1/2N) Σ_{i=1}^{N} [exp(σ_i²) - (1 + σ_i²) + μ_i²],
    # Calculate the loss value

    kl_loss = 0.5 * tf.reduce_mean(tf.exp(z_var) - (1 + z_var) + tf.square(z_mean))
    vae.add_loss(kl_loss)
    return vae
```

$$\# Loss_{u,\sigma^2} = \frac{1}{2N}\sum_{i=1}^{N}\left[exp\left(\sigma_i^2\right) - \left(1 + \sigma_i^2\right) + \mu_i^2\right],$$

The sampling layer in Fig. 10.19 is a custom network layer. Fig. 10.20 describes the output of this layer.

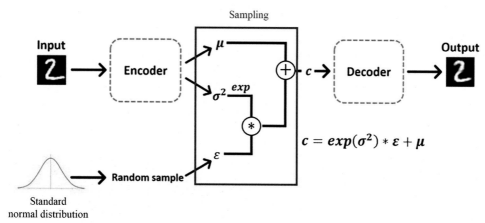

FIG. 10.20 Sampling in the Variational Auto-Encoder (VAE) model.

- The source code of the sampling custom network layer

```
class Sampling(keras.layers.Layer):
    def call(self, inputs):
        z_mean, z_log_var = inputs
        batch = tf.shape(z_mean)[0]
        dim = tf.shape(z_mean)[1]
        epsilon = tf.random.normal(shape=(batch, dim))
        return z_mean + tf.exp(z_log_var) * epsilon
```

Supplementary explanation

$Loss_{u, \sigma^2}$ (KL Loss) is used to optimize μ and σ^2 in the middle of the VAE model, so it is necessary to declare the internal loss together at the establishment of the network layer. To accomplish this task, the "vae.add_loss" used in "create_vae_model" function above is one of the methods. We can also create a "custom network model" or "custom network layer" to add the $Loss_{u, \sigma^2}$ loss function.

Example 1: custom network model

```
class VariationalAutoEncoder(keras.Model):
    def __init__(self, name='autoencoder', **kwargs):
        super(VariationalAutoEncoder, self).__init__(name=name, **kwargs)
        self.encoder = Encoder()
        self.decoder = Decoder()
        self.sampling = Sampling()

    def call(self, inputs):
        z_mean, z_var = self.encoder(inputs)
        z = self.sampling([z_mean, z_var])
        img_output = self.decoder(z)
        kl_loss=0.5*tf.reduce_mean(tf.exp(z_var)-(1+z_var)+tf.square(z_mean))
        self.add_loss(kl_loss)
        return reconstructed
```

Example 2: Custom network layer

```
class KLLoss(keras.layers.Layer):
    def call(self, inputs):
        z_mean = inputs[0]
        z_var = inputs[1]
        kl_loss=0.5*tf.reduce_mean(tf.exp(z_var)-(1+z_var)+tf.square(z_mean))
        self.add_loss(kl_loss)
        return z_mean, z_var
# ... Omit the convolution layer and fully connected layer of the Encoder ...
z_mean = keras.layers.Dense(latent_dim)(x)
z_var = keras.layers.Dense(latent_dim)(x)
z_mean, z_var = KLLoss()([z_mean, z_var])
z = Sampling()([z_mean, z_var])
encoder = keras.Model(inputs=img_inputs, outputs=z, name='encoder')
```

- VAE loss function

$$Loss_{reconstruction} = \frac{1}{N} \sum_{i=1}^{N} \sum_{x=1}^{W} \sum_{y=1}^{H} \sum_{c=1}^{C} binary\_crossentropy\left(x_{x,y,c}, \hat{y}_{x,y,c}\right)$$

x: input image.

ŷ: output image.

W: width of image.

H: height of image.

C: the number of image channels.

N: the amount of data in a batch.

The reconstruction loss function is written in the "losses.py" file.

```
def reconstruction_loss(y_true, y_pred):
    # Binary Cross-Entropy loss is used for calculating error between each pixel of the
    generated image and the input image
    bce = -(y_true * tf.math.log(y_pred + 1e-07) +
        (1 - y_true) * tf.math.log(1 - y_pred + 1e-07))
    return tf.reduce_mean(tf.reduce_sum(bce, axis=[1, 2, 3]))
```

- Custom callback

  Please write the source code in "callbacks.py" file.

  - "SaveDecoderModel" class: Check every epoch. If there is any improvement of loss, save the decoder model (similar to keras.callbacks.ModelCheckpoint).
  - "SaveDecoderOutput" class: In each epoch, the decoder model generates 225 images and write them into the TensorBoard log file for observing the output changes.

  SaveDecoderModel

```
class SaveDecoderModel(tf.keras.callbacks.Callback):
    def __init__(self, weights_file, monitor='loss', save_weights_only=False):
        super(SaveDecoderModel, self).__init__()
        self.weights_file = weights_file    # Decoder model storage path
        self.best = np.Inf   # Set best to infinite
        self.monitor = monitor
        self.save_weights_only = save_weights_only   # Save model weights

    def on_epoch_end(self, epoch, logs=None):
        """
        Each epoch, if there is an improvement of loss, the model or model weights will be
        saved
        """
        loss = logs.get(self.monitor)   # Get the value to be measured
        if loss < self.best:
            if self.save_weights_only:
                # Save the weight of the Decoder model
                self.model.get_layer('decoder').save_weights(self.weights_file)
            else:
                # Save the complete Decoder model
                self.model.get_layer('decoder').save(self.weights_file)
            self.best = loss
```

SaveDecoderOutput

```python
class SaveDecoderOutput(tf.keras.callbacks.Callback):
    def __init__(self, image_size, log_dir):
        super(SaveDecoderOutput, self).__init__()
        self.size = image_size
        self.log_dir = log_dir  # the storage path of Tensorboard log file
        n = 15  # for generating (15x15) images
        self.save_images = np.zeros((image_size * n, image_size * n, 1))
        self.grid_x = np.linspace(-1.5, 1.5, n)
        self.grid_y = np.linspace(-1.5, 1.5, n)

    def on_train_begin(self, logs=None):
        """ Tensorboard log file is created before starting training """
        path = os.path.join(self.log_dir, 'images')
        self.writer = tf.summary.create_file_writer(path)

    def on_epoch_end(self, epoch, logs=None):
        """
        225 images are generated and write into the log file
        """
        for i, yi in enumerate(self.grid_x):
            for j, xi in enumerate(self.grid_y):
                # Generate a set of Code
                z_sample = np.array([[xi, yi]])
                # Decoder generates images
                img = self.model.get_layer('decoder')(z_sample)
                # Save image
                self.save_images[i*self.size:(i+1)*self.size,
                            j*self.size:(j+1)*self.size] = img.numpy()[0]
        # write the generated 225 images to the TensorBoard log file
        with self.writer.as_default():
            tf.summary.image("Decoder output", [self.save_images], step=epoch)
```

(b) Building and training the VAE model

The source code for training the VAE model is written in a "train.py" file.

- Import packages

```python
import os
import tensorflow as tf
import tensorflow_datasets as tfds
from tensorflow import keras
from utils.models import create_vae_model
from utils.losses import reconstruction_loss
from utils.callbacks import SaveDecoderOutput, SaveDecoderModel
```

- Preparing data
  - Data normalization

```python
def parse_fn(dataset, input_size=(28, 28)):
    x = tf.cast(dataset['image'], tf.float32)
    # Resize the image to the network input size
    x = tf.image.resize(x, input_size)
    # Normalize the image
    x = x / 255.
    # Return training data and the answers
    return x, x
```

- Load MNIST dataset

```python
train_data = tfds.load('mnist', split=tfds.Split.TRAIN)
test_data = tfds.load('mnist', split=tfds.Split.TEST)
```

- Set data

```python
AUTOTUNE = tf.data.experimental.AUTOTUNE  # Automatic adjustment mode
batch_size = 16  # batch size
train_num = info.splits['train'].num_examples  # Number of training data

# Shuffle training data
train_data = train_data.shuffle(train_num)
# Training data
train_data = train_data.map(parse_fn, num_parallel_calls=AUTOTUNE)
# Set the batch size to 16 and turn on prefetch mode
train_data = train_data.batch(batch_size).prefetch(buffer_size=AUTOTUNE)

# Test data
test_data = test_data.map(parse_fn, num_parallel_calls=AUTOTUNE)
# Set the batch size to 16 and turn on prefetch mode
test_data = test_data.batch(batch_size).prefetch(buffer_size=AUTOTUNE)
```

- Set callback

```python
# Create a directory to save model weights
log_dirs = 'logs_vae'
model_dir = log_dirs + '/models'
os.makedirs(model_dir, exist_ok=True)

# Save the training log as a TensorBoard log file
model_tb = keras.callbacks.TensorBoard(log_dir=log_dirs)
# Store the best model weights
model_sdw = SaveDecoderModel(model_dir + '/best_model.h5', monitor='val_loss')
# write the image generated by Decoder to TensorBoard log file
model_testd = SaveDecoderOutput(28, log_dir=log_dirs)
```

- Create VAE model

```
# the input size of the VAE model
input_shape = (28, 28, 1)
# Dimensional space vectors
latent_dim = 2
# Create VAE model
vae_model = create_vae_model(input_shape, latent_dim)
```

- Set the optimizer and loss function

```
optimizer = tf.keras.optimizers.RMSprop()
vae_model.compile(optimizer, loss=reconstruction_loss)
```

- Train the VAE model

```
vae_model.fit(train_data,
        epochs=20,
        validation_data=test_data,
        callbacks=[model_tb, model_sdw, model_testd])
```

(c) Visualization results

- Image generation

The output image of the VAE model is generated by the decoder network. The source code for testing model is written in a "test.py" file. Fig. 10.21 presents the test results.

- Import packages:

```
import numpy as np
import tensorflow as tf
import matplotlib.pyplot as plt
```

- Load trained model for generating images:

```
size = 28  # Output image size
n = 15  # Generate (15x15) images
save_images = np.zeros((size * n, size * n, 1))
grid_x = np.linspace(-1.5, 1.5, n)
grid_y = np.linspace(-1.5, 1.5, n)

# Load trained VAE model.
model = tf.keras.models.load_model('logs_vae/models/best_model.h5')
for i, yi in enumerate(grid_x):
    for j, xi in enumerate(grid_y):
```

```
# Generate Codes
z_sample = np.array([[xi, yi]])
# Generate images
img = model(z_sample)
# Save images for displaying
save_images[i * size: (i + 1) * size, j * size: (j + 1) * size] = img.numpy()[0]

# Display generated images
plt.imshow(save_images[..., 0], cmap='gray')
plt.show()
```

Result:

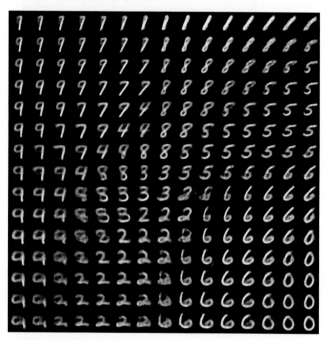

FIG. 10.21   The reconstructed images of the Varitaional Auto-Encoder (VAE) model.

- Observation of training results with TensorBoard
Open TensorBoard with command line:

```
tensorboard --logdir logs-vae
```

The prediction changes of the VAE model during training can be observed through TensorBoard, as shown in Fig. 10.22. Note the recoded image in Fig. 10.22 obtained by using the custom callback "SaveDecoderOutput."

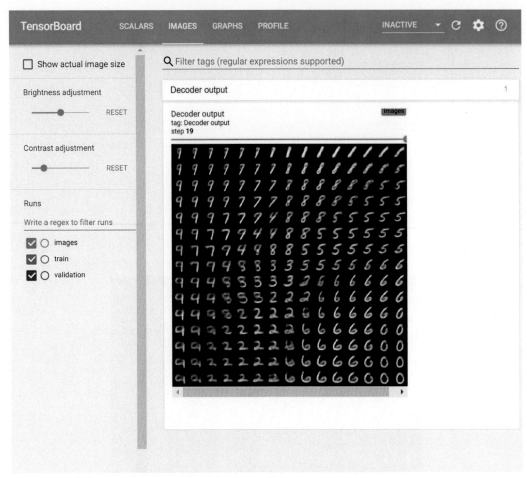

FIG. 10.22   Training results of the Variational Auto-Encoder (VAE) model on TensorBoard.

References

[1] G. Hinton, R. Salakhutdinov, Reducing the dimensionality of data with neural networks, Science 313 (5786) (2006) 504–507.

[2] G.E. Hinton, R. Zemel, Autoencoders, minimum description length and helmholtz free energy, Adv. Neural Inf. Proces. Syst. 6 (1994) 3–10.

[3] P. Vincent, et al., Stacked denoising autoencoders: learning useful representations in a deep network with a local denoising criterion, J. Mach. Learn. Res. 11 (12) (2010).

[4] D.P. Kingma, M. Welling, Auto-encoding variational bayes, in: 2nd International Conference on Learning Representations, Canada, April 14–16, 2014 [Online]. Available: http://arxiv.org/abs/1312.6114.

[5] L. Deng, The MNIST Database of handwritten digit images for machine learning research [Best of the Web], IEEE Signal Process. Mag. 29 (6) (2012) 141–142.

11

Generative adversarial network

11.1 Generative adversarial network

11.1.1 Introduction to generative adversarial network

Goodfellow et al. [1] proposed the Generative Adversarial Network (GAN) in 2014. A GAN learns to produce new data having the same statistics as the training data and consists of two subnetworks including a generator and discriminator, as shown in Fig. 11.1. The generator is responsible for generating fake samples that are indistinguishable from real samples, while the discriminator learns to identify the authenticity of the generated samples from the generator. The closer the generated samples are to the real samples, the higher the score the discriminator gives, and vice versa. Because the output of the discriminator goes through the sigmoid activation function, the score is in the range of 0 to 1.

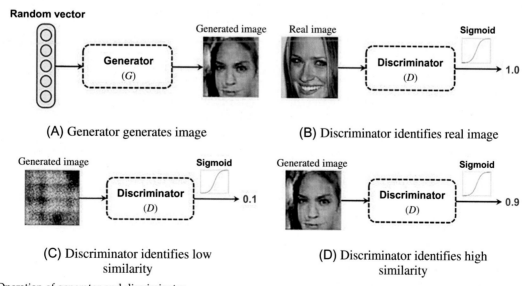

(A) Generator generates image

(B) Discriminator identifies real image

(C) Discriminator identifies low similarity

(D) Discriminator identifies high similarity

FIG. 11.1 Operation of generator and discriminator.

The operation of a GAN can be imagined as a student (generator) who learns to draw a portrait and a teacher (discriminator) who estimates and distinguishes between the portrait of the student and a real portrait. Based on the corrections of the teacher, the student learns to improve the work, forming a cycle of confrontation and improvement. At the end of the training, the portrait drawn by the student and the real portrait become indistinguishable. This is the main concept of a GAN, as shown in Fig. 11.2.

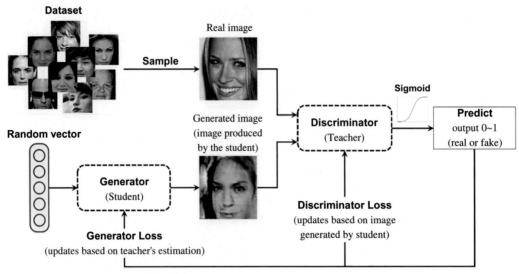

FIG. 11.2   Diagram of a Generative Adversarial Network (GAN).

Many types of GANs have been proposed, such as DCGAN [2], ImprovedGAN [3], PACGAN [4], WGAN [5], WGAN-GP [6], CycleGAN [7], PGGAN [8], StackGAN [9], video2video [10], BigGAN [11], StyleGAN [12], and so on. GANs have shown remarkable results in many computer vision tasks such as image generation, text generation, text-to-image generation, image-to-image translation, and so on, as shown in Fig. 11.3.

11.1.2 Training generative adversarial network

The training method for a GAN is different from that of network models introduced in the previous chapters. In training a GAN, two models including the generator and discriminator are simultaneously trained by an adversarial process using separate loss functions, as shown in Fig. 11.4. We introduce the loss functions for training the generator and discriminator in the following section.

1. Training generator

The purpose of training the generator is to minimize generator loss. The lesser the generator loss, the greater the ability of the generator to produce fake samples that are close to the real samples. As shown in Fig. 11.5, after inputting a set of random vectors (z) to the generator, a generated image ($\hat{x}$) is produced. Then, this resulting image is sent to the discriminator for identification. If the output of the discriminator is close to 1, the less the generator loss will be. This means that the discriminator believes that the sample generated by the generator is more realistic or, in other words, the discriminator is unable to distinguish the real sample and the generated sample (fake sample). During training of the generator, the weights of the discriminator need to be fixed. The loss function of the generator is formulated as:

$$Generator\ Loss = \frac{1}{N}\sum_{i=1}^{N} log(1 - \boxed{D(G(z^i))})$$

The closer to the 1, the lower Generator loss

G: generator.
D: discriminator.
z: input data of the generator.
N: amount of training data.

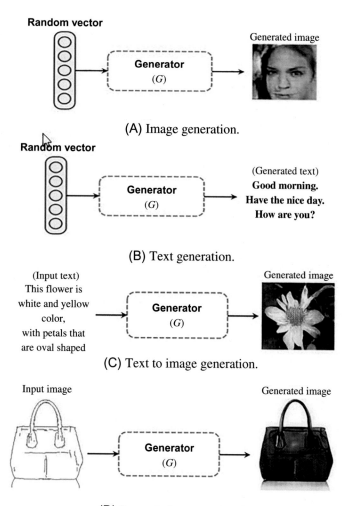

(A) Image generation.

(B) Text generation.

(C) Text to image generation.

(D) Image to image generation.

FIG. 11.3   Applications of Generative Adversarial Networks (GANs).

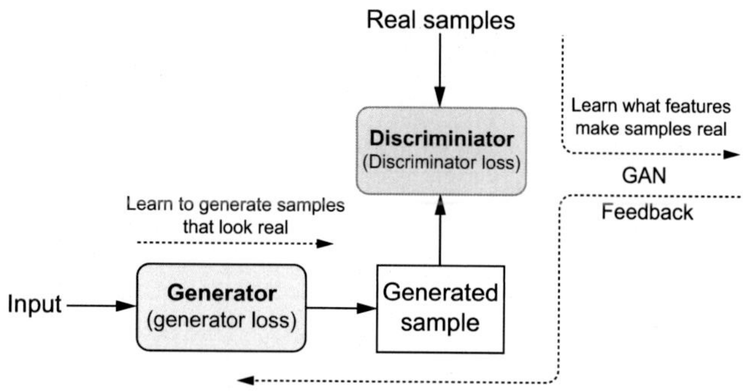

FIG. 11.4   The training process of a Generative Adversarial Network (GAN).

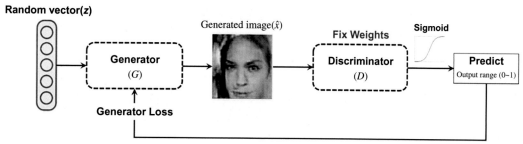

FIG. 11.5   Diagram of training the generator.

2. Training discriminator

The purpose of training a discriminator is to minimize discriminator loss. The lesser the discriminator loss, the more the discriminator is capable of distinguishing between generated samples and real samples. As shown in Fig. 11.6, after the real image (x) is inputted to the discriminator, the output prediction result is expected to be 1. After the generated image ($\hat{x}$) is produced through the generator using a set of random vectors (z) as input, it is passed through the discriminator for identification, and the output is expected to be 0. When the output of the discriminator with the real sample is closer to 1, and that of the generated sample is closer to 0, the discriminator loss is lower or, in other words, the discriminator can easily distinguish the real sample from the generated sample. During training of the discriminator, the weights of the generator need to be fixed. The objective function of the discriminator is described as:

$$\text{Discriminator Loss} = -\frac{1}{N}\sum_{i=1}^{N} log\,\boxed{D(x^i)} - \frac{1}{N}\sum_{i=1}^{N} log(1 - \boxed{D(\hat{x}^i)})$$

The closer to 0, the lower Discriminator Loss

The closer to 1, the lower Discriminator Loss

D: discriminator.
x: real image from dataset.
$\hat{x}$: generated image from the generator.
N: amount of training data.

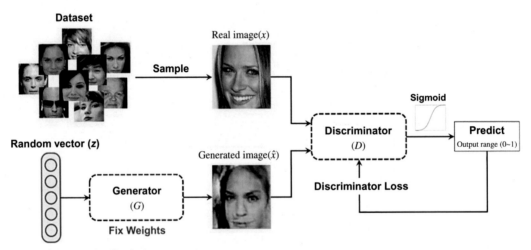

FIG. 11.6 Diagram of training the discriminator.

During training a GAN, after each iteration, the generator produces more realistic samples, while the discriminator becomes better at telling them apart, as shown in Fig. 11.7. The process reaches equilibrium when the discriminator can no longer distinguish real samples from generated samples.

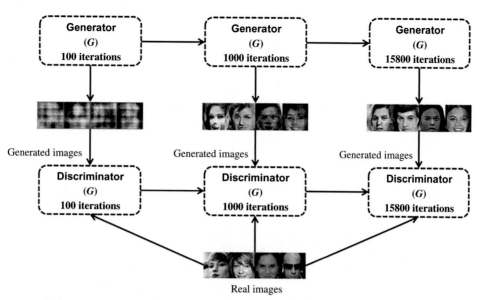

FIG. 11.7 The output result changes of the generator after the different number of training iterations.

11.2 Introduction to WGAN-GP

In this section, we analyze the drawbacks of the original GAN [1] and the Wasserstein GAN (WGAN) [5]. Then, we introduce an improved version of WGAN, the Wasserstein Generative Adversarial Network with Gradient Penalty (WGAN-GP) [6], which applies a gradient penalty technique for stable training of GANs.

11.2.1 Drawbacks of generative adversarial network

The biggest drawback of the original GAN is the instability of training, which is due to the following:

- The discriminator is trained too well, which means that the discriminator can easily distinguish real data distribution ($\mathbb{P}_{data}$) from generated data distribution ($\mathbb{P}_G$), as shown in Fig. 11.8A. $\mathbb{P}_G$ is predicted by the discriminator, and after passing through the sigmoid activation function, the output value is close to 0. The discriminator predicts $\mathbb{P}_{data}$, and after sending to the sigmoid activation function, its output value is close to 1. However, as shown in Table 5.1 in Chapter 5, when the output value of sigmoid is close to 0 or 1, its derivative approaches 0. Because of the problem of the vanishing gradient, it is difficult for the generator to update weights during the training process.
- The discriminator is not well trained, which means it cannot correctly judge between the $\mathbb{P}_{data}$ and $\mathbb{P}_G$. This may cause $\mathbb{P}_{data}$ and $\mathbb{P}_G$ to be predicted similarly. As shown in Fig. 11.8B, the $\mathbb{P}_{data}$ and $\mathbb{P}_G$ are overlaps, which means the discriminator cannot distinguish $\mathbb{P}_{data}$ from $\mathbb{P}_G$. The incorrect judgment of the discriminator also causes the generator to update in the wrong direction.

To avoid the problem shown in Fig. 11.8A and B, and achieve a well-trained model like Fig. 11.8C, discriminator loss needs to be carefully monitored during training so that there is a chance for the generator to produce the results that are closer to the real data.

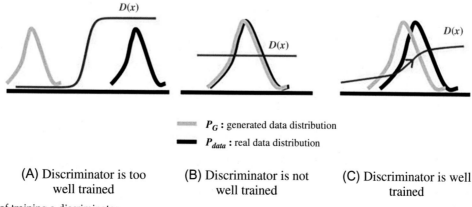

FIG. 11.8 Cases of training a discriminator.

The loss function of GAN [1] is:

$$\min_G \max_D V(G, D) = \mathbb{E}_{x \sim \mathbb{P}_{data}}\big[\log(D(x))\big] + \mathbb{E}_{x \sim \mathbb{P}_G}\big[\log(1 - D(x))\big] \tag{11.1}$$

For a fixed generator, the optimal discriminator can be obtained by:

$$D^*(x) = \frac{\mathbb{P}_{data}(x)}{\mathbb{P}_{data}(x) + \mathbb{P}_G(x)} \tag{11.2}$$

Therefore, Eq. (11.1) can be rewritten as:

$$\min_G \max_D V(G, D) = V(G, D^*) \tag{11.3}$$

$$= \mathbb{E}_{x \sim \mathbb{P}_{data}}\left[\log \frac{\mathbb{P}_{data}(x)}{\mathbb{P}_{data}(x) + \mathbb{P}_G(x)}\right] + \mathbb{E}_{x \sim \mathbb{P}_G}\left[\log\left(\frac{\mathbb{P}_G(x)}{\mathbb{P}_{data}(x) + \mathbb{P}_G(x)}\right)\right]$$

$$= -2\log 2 - KL\left(\mathbb{P}_{data}\left\|\frac{\mathbb{P}_{data} + \mathbb{P}_G}{2}\right.\right) + KL\left(\mathbb{P}_G\left\|\frac{\mathbb{P}_{data} + \mathbb{P}_G}{2}\right.\right)$$

$$= -2\log 2 + 2JS(\mathbb{P}_{data}\|\mathbb{P}_G)$$

$\mathbb{P}_{data}$: real data distribution.
$\mathbb{P}_G$: generated data distribution.
D: discriminator.
D^*: optimized discriminator.

KL: Kullback–Leibler divergence, also known as relative entropy, which is used to measure the difference between two probability distributions $\mathbb{P}$ and $\mathbb{Q}$ on the same probability space X, defined as:

$$KL(\mathbb{P}\|\mathbb{Q}) = \sum_{x \in X} \mathbb{P}(x) \log \frac{\mathbb{P}(x)}{\mathbb{Q}(x)} \tag{11.4}$$

JS: Jensen–Shannon divergence (JS divergence), which is based on Kullback–Leibler divergence to measure the similarity between two probabilities $\mathbb{P}$ and $\mathbb{Q}$, defined as:

$$JS(\mathbb{P}\|\mathbb{Q}) = \frac{1}{2}KL\left(\mathbb{P}\left\|\frac{\mathbb{P} + \mathbb{Q}}{2}\right.\right) + \frac{1}{2}KL\left(\mathbb{Q}\left\|\frac{\mathbb{P} + \mathbb{Q}}{2}\right.\right) \tag{11.5}$$

In Eq. (11.3), JS divergence is employed, so when minimizing the GAN loss function with an optimal discriminator, it is referred to as minimizing the JS divergence. If $\mathbb{P}_{data} = \mathbb{P}_G$, then $JS(\mathbb{P}\|\mathbb{Q}) = 0$, which means the training model achieves the global minimum with the value of -2log2, and the generator perfectly replicating the data distribution.

11.2.2 Introduction to Wasserstein distance

As mentioned, training a GAN is hard and unstable. To address this problem, the Wasserstein GAN (WGAN) [5] proposed a new loss function that uses Wasserstein distance, which has a smoother gradient everywhere. The Wasserstein distance refers to the minimum cost of transporting mass when moving one data distribution to another data distribution. For the real data distribution ($\mathbb{P}_{data}$) and generated data distribution ($\mathbb{P}_G$) by the generator, Wasserstein distance is:

$$W(\mathbb{P}_{data}, \mathbb{P}_G) = \inf_{\gamma \sim \Pi(\mathbb{P}_{data}, \mathbb{P}_G)} \mathbb{E}_{(x, y) \sim \gamma}[\|x - y\|] \tag{11.6}$$

$\Pi(\mathbb{P}_{data}, \mathbb{P}_G)$: a set of all possible transport plan γ, where joint distributions $\gamma(x, y)$ whose marginals are respectively $\mathbb{P}_{data}$ and $\mathbb{P}_G$.

$\|x - y\|$: distance between x and y
$\mathbb{E}$: expected moving cost of the transport plan γ
inf: infimum or the greatest lower bound for any transport plan γ

According to the Kantorovich–Rubinstein duality [13], Eq. (11.6) of Wasserstein distance can be converted into the following form:

$$W(\mathbb{P}_{data}, \mathbb{P}_G) = \max_{D \in 1-Lipschitz} \big\{\mathbb{E}_{x \sim \mathbb{P}_{data}}[D(x)] - \mathbb{E}_{x \sim \mathbb{P}_G}[D(x)]\big\} \tag{11.7}$$

$\mathbb{P}_{data}$: real data distribution
$\mathbb{P}_G$: Generated data distribution
D: the discriminator of the WGAN,
1-Lipschitz: for enforcing a Lipschitz constraint ($\|D(x_1) - D(x_2)\| \leq \|x_1 - x_2\|$) on the discriminator

Supplementary explanation

Unlike in a GAN [1], the discriminator in a WGAN [5] does not employ the sigmoid activation function at the output layer, and it outputs a scalar score. If the discriminator does not satisfy the 1-Lipschitz function, its output with input samples from $\mathbb{P}_{data}$ can approach $+\infty$ and its output with input samples from $\mathbb{P}_G$ can approach $-\infty$, resulting in crashing the training, as shown in Fig. 11.9.

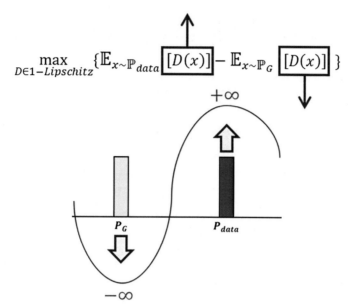

FIG. 11.9   The output of the discriminator.

To enforce the Lipschitz constraint on the discriminator, the WGAN uses a simple weight-clipping method to restrict the weights of the discriminator to be within a certain range that is controlled by a clipping threshold c. In particular, if the weights are greater than c, they are set to c; if the weights are less than −c, they are set to −c; otherwise, the weights are kept unchanged.

However, using the weight-clipping method to force the Lipschitz constraint on the discriminator encounters several problems:

- The work in [6] shows that when the WGAN is trained on the Swiss Roll dataset, and the weights of the discriminator are limited to a certain range [−0.01, 0.01], the weight distribution of the discriminator is not even. Most of the weights are concentrated in the maximum and minimum values of the clipping range, which makes the model unable to describe complex problems.
- The model performance is very sensitive to the clipping threshold, so it is difficult to tune the clipping threshold c correctly [6]. If c is tuned slightly larger, the gradient becomes larger every time when passing through the network layers; this leads to a gradient explosion problem. On the contrary, if c is tuned slightly smaller, the gradient becomes smaller every time when passing through the network layers; this leads to gradient vanishing after multiple transmissions.

To solve the problems of the WGAN, the WGAN-GP [6] is proposed, which uses the gradient penalty method instead of the weight-clipping method to enforce the 1-Lipschitz constraint on the discriminator. In the WGAN-GP, the gradient penalty technique is applied by directly constraining the gradient norm of the output of the discriminator with respect to its input. The objective function of the WGAN-GP is:

$$L = \max_{D}\{\mathbb{E}_{x\sim\mathbb{P}_{data}}[D(x)] - \mathbb{E}_{x\sim\mathbb{P}_G}[D(x)] - \boxed{\lambda\mathbb{E}_{x\sim\mathbb{P}_{penalty}}[(\|\nabla_x D(x)\|_2 - 1)^2]}\} \qquad \overset{\text{Grendient penalty}}{} \tag{11.8}$$

$\mathbb{P}_{data}$: real data distribution

$\mathbb{P}_G$: generated data distribution

$\mathbb{P}_{penalty}$: sampling uniformly along straight lines between pairs of points sampled from $\mathbb{P}_{data}$ and $\mathbb{P}_G$

D: discriminator

λ: penalty coefficient; this parameter is set to 10 as default in the original work [6]

11.2.3 Training WGAN-GP

The training process of the WGAN-GP is the same as that of the original GAN, but both the loss functions of the generator and discriminator are changed for performance improvement.

- The purpose of training the generator is to pursue the least generator loss. The lesser the generator loss is, the greater is the ability of the generator to produce samples that are close to the real samples. For example, after a set of random vectors (z) is inputted into the generator, an image ($\hat{x}$) is produced. Then, this resulting image is sent to the discriminator for prediction. The greater the output prediction of the discriminator, the lesser the generator loss becomes. During training of the generator, the weights of the discriminator need to be fixed. The objective function of the generator is:

$$\overset{\text{The bigger value refers to lower loss}}{Generator\ Loss = -\frac{1}{N}\sum_{i=1}^{N}\boxed{D(G(z^i))}} \tag{11.9}$$

D: discriminator

G: generator

z: input sample of the generator

N: amount of training data

- The purpose of training the discriminator is to pursue the least discriminator loss. The lesser the discriminator loss is, the better the discriminator's ability to distinguish between generated samples and real samples. During training of the discriminator, the weights of the generator need to be fixed. The loss function of the discriminator is formulated as:

$$Discriminator\ Loss = -\frac{1}{N}\sum_{i=1}^{N}\boxed{D(x^i)} + \frac{1}{N}\sum_{i=1}^{N}\boxed{D(\hat{x}^i)} + \lambda \times Gradient\ Penalty \tag{11.10}$$

The smaller value refers to lower loss

The bigger value refers to lower loss

D: discriminator

x: real sample from dataset

$\hat{x}$: generated sample from the generator

λ: penalty coefficient; usually set to 10

N: amount of training data

- Gradient penalty: considering directly constraining the gradient norm of the output of the discriminator with respect to its input. The formula for gradient penalty is:

The closer to 1, the lower the loss value

$$Gradient\ Penalty = \frac{1}{N}\sum_{i=1}^{N}\boxed{(\|\nabla_{\tilde{x}}D(\tilde{x})\|_2}-1)^2 \tag{11.11}$$

$$\tilde{x} = tx + (1-t)\hat{x}$$

D: discriminator
$\tilde{x}$: sampled from the real sample (x) and the generated sample ($\hat{x}$) with t uniformly sampled between 0 and 1
N: amount of training data

11.3 Experiment: Implementation of WGAN-GP

This section introduces how to build the WGAN-GP model for image generation. To conduct experiments, we used the large-scale CelebFaces Attributes (CelebA) dataset [14] to train and test the model. Fig. 11.10 shows some generated images of the WGAN-GP after different training iterations.

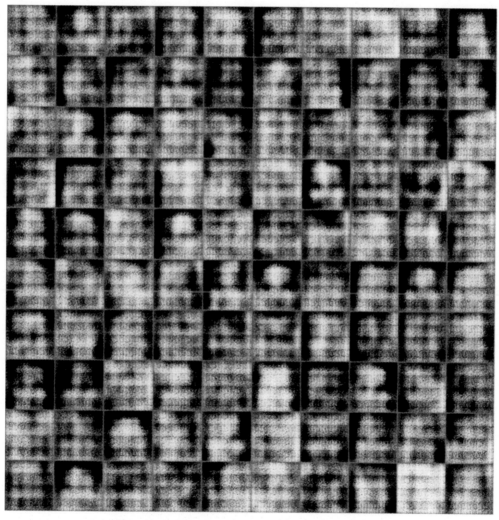

(A) Generated images after 100 training iterations

FIG. 11.10

(Continued)

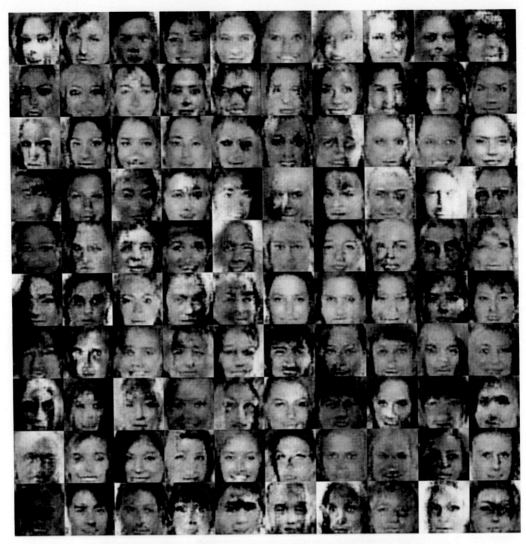

(B) Generated images after 1000 training iterations

FIG. 11.10, Cont'd

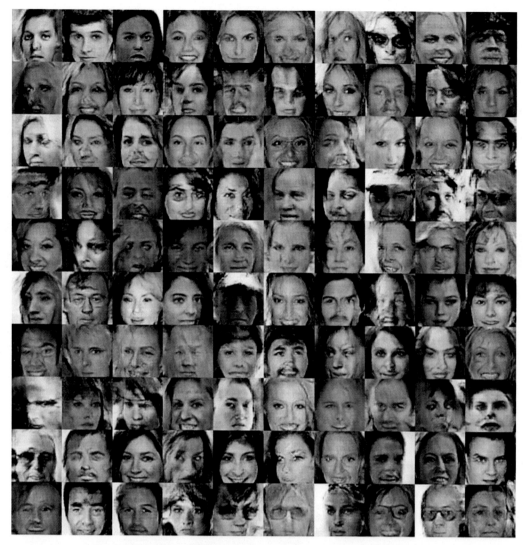

(C) Generated images after 15800 training iterations

FIG. 11.10 Generated images of the WGAN-GP after different training iterations.

11.3.1 Create project

Since the example program of WGAN-GP in this chapter is more complicated than that of models in Chapters 1–9, we employ PyCharm IDE as a compiler to write the source codes and train the models. In the following, we outline the process of creating the project.

1. Create a new project: Click "File" → "New Project," as shown in Fig. 11.11.

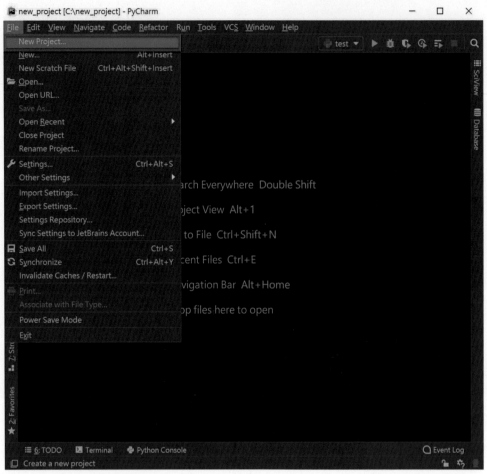

FIG. 11.11　Creating a new project on PyCharm.

2. Set the directory of the new project, as shown in Fig. 11.12.

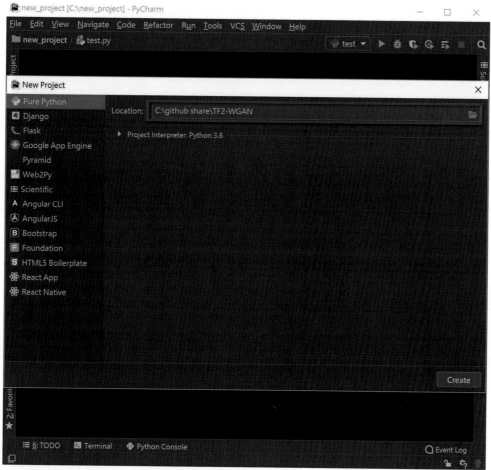

FIG. 11.12 Setting a new project directory.

3. Configure a Python interpreter: Open "Project Interpreter: Python 3.6," select the "Existing interpreter," and set Python Interpreter, as shown in Fig. 11.13.

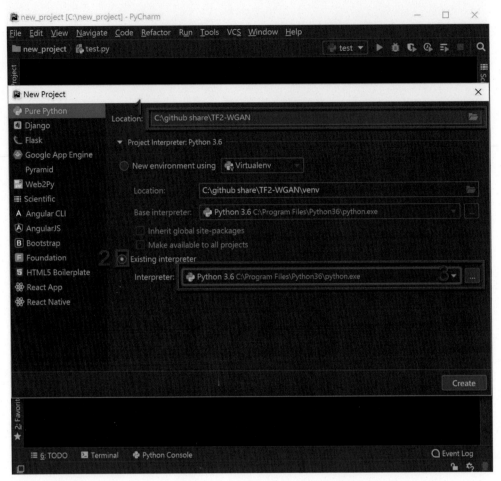

FIG. 11.13   Interpreter setting.

4. Create a project: Click button "Create" to create a new project, as shown in Fig. 11.14.

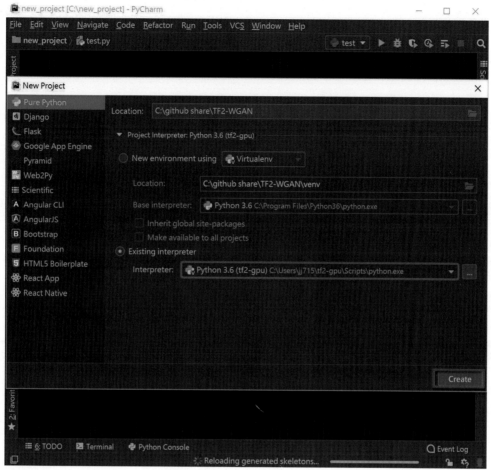

FIG. 11.14 Create a project.

Supplementary explanation

The source code for the WGAN-GP in this chapter can be downloaded at https://github.com/taipeitechmmslab/MMSLAB-DL/tree/master, as shown in Fig. 11.15.

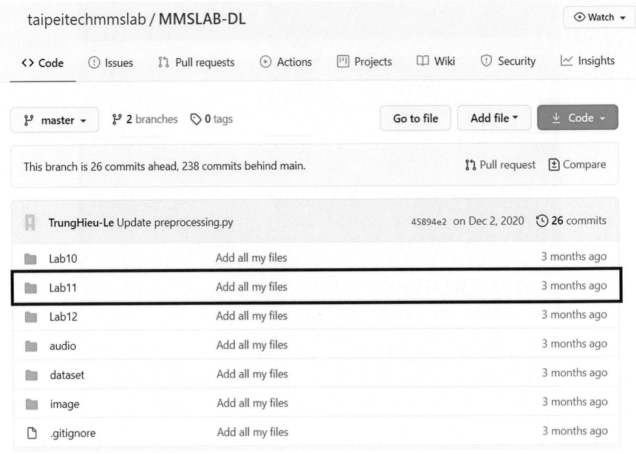

FIG. 11.15   The source code for WGAN-GP on GitHub.

11.3.2 Introduction to dataset

The CelebA dataset [14] contains 202,599 images of size $218 \times 178 \times 3$, divided into three sets, including 162,770 images in the training set, 19,962 images in the verification set, and 19,867 images in the test set. Fig. 11.16 shows some example images from the CelebA dataset. The dataset can be loaded through TensorFlow datasets as follows.

```
import tensorflow_datasets as tfds
# Load training set
train_data, info = tfds.load("celeb_a", split= tfds.Split.TRAIN, with_info=True)
# Load validation set
valid_data = tfds.load("celeb_a", split= tfds.Split.VALIDATION)
# # Load test set
test_data = tfds.load("celeb_a", split= tfds.Split.TEST)
```

FIG. 11.16 Images from CelebA dataset.

11.3.3 Building WGAN-GP model

1. Directory and files

The Python files of the WGAN-GP project are shown in Fig. 11.17. The files are as follows.

- train.py: file source code for training WGAN-GP model
- utils:
  - models.py: file source code for WGAN-GP model
  - losses.py: file source code for custom loss function
  - dataset.py: file source code for data preprocessing

FIG. 11.17 WGAN-GP project.

2. Implementing the WGAN-GP model

Fig. 11.18 shows a flowchart of the source code for building the WGAN-GP model.

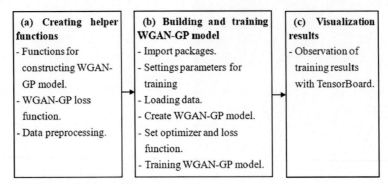

FIG. 11.18 Flowchart of the source code for the WGAN-GP model.

(a) Creating helper functions
- Functions for constructing WGAN-GP model

The network architecture of WGAN-GP is defined in the "models.py" file. Fig. 11.19 shows the architecture of the generator and discriminator.

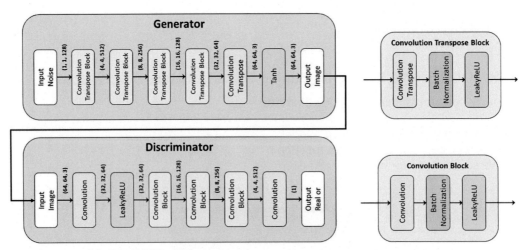

FIG. 11.19 The architecture of the WGAN-GP model.

The functions for creating the generator and discriminator.

```python
def Generator(input_shape=(1, 1, 128), name='Generator'):
    inputs = keras.Input(shape=input_shape)

    # 1: Convolution Transpose Block1, 1x1 -> 4x4
    x = keras.layers.Conv2DTranspose(512, 4, strides=1,
                    padding='valid', use_bias=False)(inputs)
    x = keras.layers.BatchNormalization()(x)
    x = keras.layers.LeakyReLU()(x)
    # 2: Convolution Transpose Block2, 4x4 -> 8x8
    x = keras.layers.Conv2DTranspose(256, 4, strides=2,
                    padding='same', use_bias=False)(x)
    x = keras.layers.BatchNormalization()(x)
    x = keras.layers.LeakyReLU()(x)
    # 3: Convolution Transpose Block3, 8x8 -> 16x16
    x = keras.layers.Conv2DTranspose(128, 4, strides=2,
                    padding='same', use_bias=False)(x)
    x = keras.layers.BatchNormalization()(x)
    x = keras.layers.LeakyReLU()(x)
    # 4: Convolution Transpose Block4, 16x16 -> 32x32
    x = keras.layers.Conv2DTranspose(64, 4, strides=2,
                    padding='same', use_bias=False)(x)
    x = keras.layers.BatchNormalization()(x)
    x = keras.layers.LeakyReLU()(x)
    # 5: Convolution Transpose + Tanh, 32x32 -> 64x64
    x = keras.layers.Conv2DTranspose(3, 4, strides=2, padding='same', use_bias=False)(x)
    outputs = keras.layers.Activation('tanh')(x)
    return keras.Model(inputs=inputs, outputs=outputs, name=name)


def Discriminator(input_shape=(64, 64, 3), name='Discriminator'):
    inputs = keras.Input(shape=input_shape)
    # 1: Convolution + LeakyReLU, 64x64 -> 32x32
    x = keras.layers.Conv2D(64, 4, strides=2, padding='same')(inputs)
    x = keras.layers.LeakyReLU()(x)
    # 2: Convolution Block1, 32x32 -> 16x16
    x = keras.layers.Conv2D(128, 4, strides=2, padding='same', use_bias=False)(x)
    x = keras.layers.BatchNormalization()(x)
    x = keras.layers.LeakyReLU()(x)
    # 3: Convolution Block2, 16x16 -> 8x8
    x = keras.layers.Conv2D(256, 4, strides=2, padding='same', use_bias=False)(x)
    x = keras.layers.BatchNormalization()(x)
    x = keras.layers.LeakyReLU()(x)
```

```python
    # 4: Convolution Block3, 8x8 -> 4x4
    x = keras.layers.Conv2D(512, 4, strides=2, padding='same', use_bias=False)(x)
    x = keras.layers.BatchNormalization()(x)
    x = keras.layers.LeakyReLU()(x)
    # 5: Convolution, 4x4 -> 1x1
    outputs = keras.layers.Conv2D(1, 4, strides=1, padding='valid')(x)
    return keras.Model(inputs=inputs, outputs=outputs, name=name)
```

- WGAN-GP loss function
  - Generator loss: The purpose of this loss function is to train the generator to produce images that are indistinguishable from the real images when using the discriminator.

$$GeneratorLoss = -\frac{1}{N}\sum_{i=1}^{N}D\left(G\left(z^{i}\right)\right)$$

D: discriminator
G: generator
z: input sample of the generator
N: amount of training data

The generator loss function is described in the "losses.py" file.

```
def generator_loss(fake_logit):
    g_loss = - tf.reduce_mean(fake_logit)
    return g_loss
```

- Discriminator loss: Reducing the real_loss value allows the discriminator easily predict the real image, reducing the fake_loss value allows the discriminator to easily identify the fake image, and the gradient penalty allows the discriminator to satisfy the 1-Lipschitz constraint. The discriminator loss function is described in the "losses.py" file.

$$Discriminator\ Loss = \boxed{-\frac{1}{N}\sum_{i=1}^{N}D(x^{i})} + \boxed{\frac{1}{N}\sum_{i=1}^{N}D(\hat{x}^{i})} + \lambda \times Gradient\ Penalty$$

$$\underset{real\_loss}{} \qquad \underset{fake\_loss}{}$$

D: discriminator
x: real image from dataset
$\hat{x}$: generated image from the generator
λ: penalty coefficient; usually set to 10
N: amount of training data

The functions for real_loss and fake_loss

```
def discriminator_loss(real_logit, fake_logit):
    real_loss = - tf.reduce_mean(real_logit)
    fake_loss = tf.reduce_mean(fake_logit)
    return real_loss, fake_loss
```

- Gradient penalty

$$GradientPenalty = \frac{1}{N}\sum_{i=1}^{N}\left(\left\|\nabla_{\tilde{x}}D\left(\tilde{x}\right)\right\|_{2} - 1\right)^{2}$$

$$\tilde{x} = tx + (1-t)\hat{x}$$

D: discriminator
$\tilde{x}$: sampled from the real image (x) and the generated image ($\hat{x}$) with t uniformly sampled between 0 and 1
N: amount of training data

The function for gradient penalty

```
def gradient_penalty(discriminator, real_img, fake_img):
    def _interpolate(a, b):
        shape = [tf.shape(a)[0]] + [1] * (a.shape.ndims - 1)
        alpha = tf.random.uniform(shape=shape, minval=0., maxval=1.)
        inter = (alpha * a) + ((1 - alpha) * b)
        inter.set_shape(a.shape)
        return inter
    # Perform interpolation the generated image and the real image to obtain x̃
    x_img = _interpolate(real_img, fake_img)
    with tf.GradientTape() as tape:
        # ensure that x_img can be tracked by tape
        tape.watch(x_img)
        # Discriminator predict image
        pred_logit = discriminator(x_img)
```

```
    # Compute gradient
    grad = tape.gradient(pred_logit, x_img)
    # Calculate the norm of the gradient
    norm = tf.norm(tf.reshape(grad, [tf.shape(grad)[0], -1]), axis=1)
    # L2 normalization
    gp_loss = tf.reduce_mean((norm - 1.)**2)
    return gp_loss
```

- Data prepossessing

Data prepossessing: using input image size of 64×64, and each pixel of the input image is normalized to scale the value to the range $[-1,+1]$. The code is written in the "dataset.py" file.

The function of data preprosesing

```
def parse_fn(dataset, input_size=(64, 64)):
    x = tf.cast(dataset['image'], tf.float32)
    crop_size = 108
    # Image size (218, 178, 3)
    h, w, _ = x.shape
    # Crop images
    x = tf.image.crop_to_bounding_box(x, (h-crop_size)//2, (w-crop_size)//2,
                                      crop_size, crop_size)
    # Resize the image with size of (108, 108, 3) to (64, 64, 3)
    x = tf.image.resize(x, input_size)
        # Normalize the image to -1~+1，
        # Steps: [0~255]/127.5→[0~2], [0~2]-1→[-1~1]
    x = x / 127.5 - 1
    return x
```

(b) Building and training the WGAN-GP model

For this part of training the WGAN-GP model, the source code is written in the "train.py" file.

- Import necessary packages

```python
import numpy as np
import tensorflow as tf
import tensorflow_datasets as tfds
from functools import partial
from utils.dataset import parse_fn
from utils.losses import generator_loss, discriminator_loss, gradient_penalty
from utils.models import Generator, Discriminator
```

- Settings parameters for training

```python
# Batch size of 64
batch_size = 64
# learning rate is set to 1× 10⁻⁴
lr = 0.0001
# the input size of Generator
z_dim = 128
# Discriminator is trained 5 times, and Generator is trained 1 time
n_dis = 5
# Set the penalty coefficient, usually set to 10
gradient_penalty_weight = 10.0
```

- Loading data

```python
#combining data
combine_split = tfds.Split.TRAIN + tfds.Split.VALIDATION + tfds.Split.TEST
# Loading data
train_data, info = tfds.load('celeb_a', split=combine_split, with_info=True)
# Auto adjustment mode
AUTOTUNE = tf.data.experimental.AUTOTUNE
# Shuffle data
train_data = train_data.shuffle(1000)
# Training data
train_data = train_data.map(parse_fn, num_parallel_calls=AUTOTUNE)
# Set the batch size to 64,
train_data = train_data.batch(batch_size, drop_remainder=True)
# Turn on prefetch mode
train_data = train_data.prefetch(buffer_size=AUTOTUNE)
```

- Create WGAN-GP model

```python
#Generator
generator = Generator((1, 1, z_dim))
#Discriminator
discriminator = Discriminator((64, 64, 3))
```

- Set the optimizers and loss functions
  - Setting optimizers

```python
g_optimizer = tf.keras.optimizers.Adam(lr, beta_1=0.5, beta_2=0.9)
d_optimizer = tf.keras.optimizers.Adam(lr, beta_1=0.5, beta_2=0.9)
```

- Setting function for training the generator

```python
@tf.function
def train_generator():
    with tf.GradientTape() as tape:
        # Generate a 128-dimensional vector
        random_vector = tf.random.normal(shape=(batch_size, 1, 1, z_dim))
        # Generate fake image
        fake_img = generator(random_vector, training=True)
        # Use Discriminator to evaluate
        fake_logit = discriminator(fake_img, training=True)
        # Calculate Generator Loss
        g_loss = generator_loss(fake_logit)
    # Compute gradient
    gradients = tape.gradient(g_loss, generator.trainable_variables)
    # Update Generator weights
    g_optimizer.apply_gradients(zip(gradients, generator.trainable_variables))
    return g_loss
```

- Setting function for training the discriminator

```python
@tf.function
def train_discriminator(real_img):
    with tf.GradientTape() as t:
        # Generate a 128-dimensional vector
        random_vector = tf.random.normal(shape=(batch_size, 1, 1, z_dim))
        # Generate fake image
        fake_img = generator(random_vector, training=True)
        # Use Discriminator to evaluate real image
        real_logit = discriminator(real_img, training=True)
        # Use Discriminator to evaluate fake image
        fake_logit = discriminator(fake_img, training=True)
        # Calculate the loss of Discriminator
```

```python
        real_loss, fake_loss = discriminator_loss(real_logit, fake_logit)
        # Calculate Gradient Penalty
        gp_loss = gradient_penalty(partial(discriminator,training=True),
                            real_img,fake_img)
        # Calculate Discriminator Loss
        d_loss = (real_loss + fake_loss) + gp_loss * gradient_penalty_weight
    # Compute gradient
    D_grad = t.gradient(d_loss, discriminator.trainable_variables)
    # Update Discriminator weight
    d_optimizer.apply_gradients(zip(D_grad, discriminator.trainable_variables))
    return real_loss + fake_loss, gp_loss
```

- Function for displaying image:

```python
def combine_images(images, col=10, row=10):
    # to make the image display normally, the image is scaled from -1~+1 to 0~1
    images = (images + 1) / 2
    # Convert TensorFlow format to Numpy format
    images = images.numpy()
    # Get the shape of the generated image, shape=(batch size, height, width, channel)
    b, h, w, _ = images.shape
    # Create a 10x10 array to store 100 images
    images_combine = np.zeros(shape=(h*col, w*row, 3))
    # Put 100 pictures into a 10x10 array
    for y in range(col):
        for x in range(row):
            images_combine[y*h:(y+1)*h, x*w:(x+1)*w] = images[x+y*row]
    return images_combine
```

- Training WGAN-GP:

```python
def train_wgan():
    # Create a directory to save the Generator model
    log_dirs = 'logs_wgan'
    model_dir = log_dirs + '/models/'
    os.makedirs(model_dir, exist_ok=True)
    # Create TensorBoard log
    summary_writer = tf.summary.create_file_writer(log_dirs)
```

```python
    # create a set of random vector
    sample_random_vector = tf.random.normal((100, 1, 1, z_dim))
    # A total of 25 Epochs
    for epoch in range(25):
        # Read training data (real image)
        for step, real_img in enumerate(train_data):
            # Training Discriminator
            d_loss, gp = train_discriminator(real_img)
            # Write Discriminator loss value to TensorBoard log
            with summary_writer.as_default():
                tf.summary.scalar('discriminator_loss', d_loss, d_optimizer.iterations)
                tf.summary.scalar('gradient_penalty', gp, d_optimizer.iterations)


            # 5 times of training Discriminator, 1 time of training Generator.
            if d_optimizer.iterations.numpy() % n_dis == 0:
                # Training Generator
                g_loss = train_generator()
                # Save the generator loss to TensorBoard logs
                with summary_writer.as_default():
                    tf.summary.scalar('generator_loss', g_loss, g_optimizer.iterations)
                print('G Loss: {:.2f}\tD loss: {:.2f}\tGP Loss {:.2f}'.format(g_loss,
                                    d_loss, gp))


                if g_optimizer.iterations.numpy() % 100 == 0:
                    # Generate 100 images
                    x_fake = generator(sample_random_vector, training=False)
                    # Put generated 100 images into a 10x10 array for displaying
                    save_img = combine_images(x_fake)
                    # write 100 generated images to TensorBoard logs
                    with summary_writer.as_default():
                        tf.summary.image(dataset, [save_img],
                                step=g_optimizer.iterations)
        # Each Epoch save the generator model weights
        if epoch != 0:
            generator.save_weights(model_dir+"generator-epochs-{}.h5".format(epoch))
```

(c) Visualization results

The changes of the output image results of the generator can be observed through TensorBoard, as shown in Fig. 11.20. Figs. 11.21 and 11.22 display the changes of generator loss and discriminator loss, and gradient penalty, respectively.

Open TensorBoard through the command line:

```
tensorboard --logdir logs-wgan
```

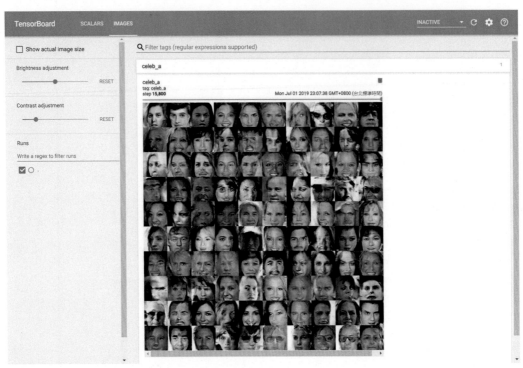

FIG. 11.20   Generated images of the generator on TensorBoard.

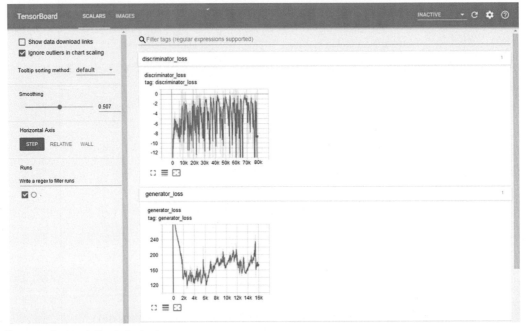

FIG. 11.21   Generator loss and discriminator loss.

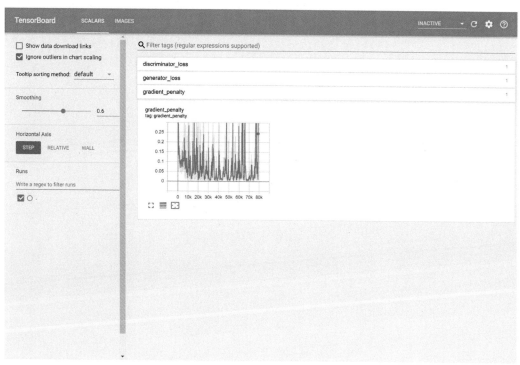

FIG. 11.22 Gradient penalty.

References

[1] I. Goodfellow, J. Pouget-Abadie, M. Mirza, B. Xu, D. Warde-Farley, S. Ozair, A. Courville, Y. Bengio, Generative adversarial nets, Adv. Neural Inf. Proces. Syst. (2014) 2672–2680.

[2] A. Radford, L. Metz, S. Chintala, Unsupervised representation learning with deep convolutional generative adversarial networks, in: International Conference on Learning Representations, 2016, pp. 1–16.

[3] T. Salimans, I. Goodfellow, W. Zaremba, V. Cheung, A. Radford, X. Chen, Improved techniques for training gans, Adv. Neural Inf. Proces. Syst. (2016) 2226–2234.

[4] Z. Lin, A. Khetan, G. Fanti, S. Oh, PacGAN: the power of two samples in generative adversarial networks, Adv. Neural Inf. Proces. Syst. (2018) 1503–1512.

[5] M. Arjovsky, S. Chintala, L. Bottou, Wasserstein GAN, in: Proc. 34th International Conference on Machine Learning, 2017, pp. 214–223.

[6] I. Gulrajani, F. Ahmed, M. Arjovsky, V. Dumoulin, A. Courville, Improved training of wasserstein GANS, Adv. Neural Inf. Proces. Syst. (2017) 5767–5777.

[7] J.-Y. Zhu, T. Park, P. Isola, A.A. Efros, Unpaired image-to-image translation using cycle-consistent adversarial networks, in: International Conference on Computer Vision, 2017, pp. 2223–2232.

[8] T. Karras, T. Aila, S. Laine, J. Lehtinen, Progressive growing of GANs for improved quality, stability, and variation, in: International Conference on Learning Representations, 2018, pp. 1–26.

[9] H. Zhang, T. Xu, H. Li, S. Zhang, X. Huang, X. Wang, D. Metaxas, Stackgan: text to photo-realistic image synthesis with stacked generative adversarial networks, in: Proceedings of the IEEE International Conference on Computer Vision, 2017, pp. 5907–5915.

[10] T.-C. Wang, M.-Y. Liu, J.-Y. Zhu, G. Liu, A. Tao, J. Kautz, B. Catanzaro, Video-to-video synthesis, in: Advances in Neural Information Processing Systems, 2018.

[11] A. Brock, J. Donahue, K. Simonyan, Large scale GAN training for high fidelity natural image synthesis, in: 7th International Conference on Learning Representations, ICLR 2019, New Orleans, LA, USA, May 6–9, 2019 [Online]. Available: https://openreview.net/forum?id=B1xsqj09Fm.

[12] T. Karras, S. Laine, T. Aila, A style-based generator architecture for generative adversarial networks, in: Proceedings of the IEEE Conference on Computer Vision and Pattern Recognition, 2019, pp. 4401–4410.

[13] C. Villani, Optimal Transport: Old and New, vol. 338, Springer Science & Business Media, 2008.

[14] Z. Liu, P. Luo, X. Wang, X. Tang, Large-scale celebfaces attributes (celeba) dataset. Retrieved August 15, 2018.

12

Object detection

OUTLINE
- Introduction to computer vision
- Introduction to state-of-the-art CNN-based object detection methods
- Implementing the YOLO-v3 object detection model

12.1 Computer vision

Computer vision is an interdisciplinary scientific field that focuses on addressing how computers can understand the contents of images or videos for automating tasks like the human visual system. Computer vision has a wide range of applications such as video surveillance [1–3], face recognition [4, 5], face annotation [6], image retrieval [7, 8], biometrics [9, 10], traffic monitoring [11, 12], visibility restoration [13–19], object detection [20–33], and so on. The four main tasks of computer vision are:

- Image classification: To identify the category of the image, as shown in Fig. 12.1A. The common network models used for image classification include GoogLeNet [34], ResNet [35], and so on.
- Object detection: To find out the category of objects and their location in a given image, as shown in Fig. 12.1B. The common network models used for object detection include Faster R-CNN [23], YOLO-v3 [30], and so on.
- Semantic segmentation: To assign a label to every pixel in the image, as shown in Fig. 12.1C. The common network models used for semantic segmentation include FCN [36], DeepLab-v3 [37], and so on.
- Instance segmentation: To identify the boundary of each object at the detailed pixel level in the image, as shown in Fig. 12.1D. The common network models used for instance segmentation include FCIS [38], Mask R-CNN [29], and so on.

Because object detection is widely applied in various areas of computer vision, this chapter focuses on introducing object detection methods, especially state-of-the-art CNN-based models.

12.2 Introduction to object detection

Object detection is one of the most important tasks in computer vision. It not only classifies the category of objects but also determines the location of them on the image. As shown in Fig. 12.2, the network takes an image as input, and then predicts multiple classes and bounding boxes for this image input. Finally, it frames and marks the label for each predicted object on the output image.

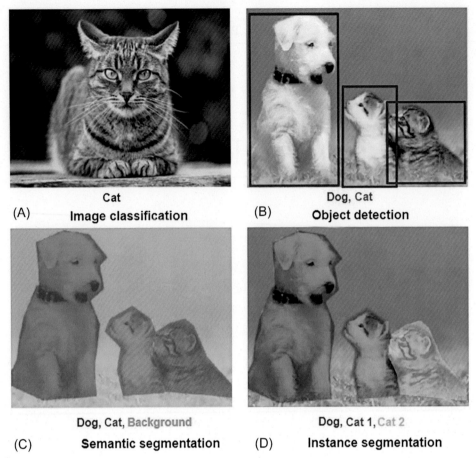

FIG. 12.1   Four main tasks in computer vision.

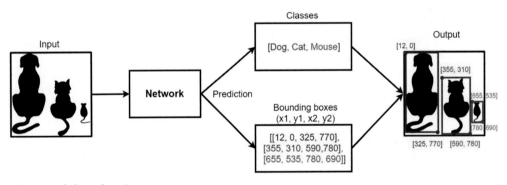

FIG. 12.2   Block diagram of object detection.

Supplementary explanation

There are many bounding box formats, such as (x1, y1, x2, y2) shown in Fig. 12.2, where (x1, y1) and (x2, y2) are the top-left and bottom-right coordinates of the object on the image, respectively. Another common format like (x, y, w, h), where (x, y) is the top-left coordinate, w and h are the width and height of the bounding box, respectively. Each object detection method can employ different bounding box formats for object location prediction.

Object detection based on convolutional neural network (CNN) has received considerable attention and achieved impressive performance in recent years. It can be divided into two groups [39]: (1) region proposal-based approach and (2) regression/classification-based approach.

- Region proposal-based approach: Following this approach, the object detection methods perform prediction in two steps. In the first step, the regions of interest (RoIs) of objects in a given image are created by using region proposal methods such as selective search [40]. In the second step, the RoIs are fed into a CNN for classifying objects and refining the bounding boxes. Although the running time of the region proposal-based approaches might be relatively slow because of predicting objects in two separate stages, they usually obtain high performance in detecting objects. The representative network models of the first group include R-CNN [20], Fast R-CNN [22], Faster R-CNN [23], and so on.
- Regression/classification-based approach: The object detection methods belonging to this group use only one step for predicting the object, mapping directly from the whole input image to bounding boxes and class probabilities. These methods can achieve real-time speed, making them applicable to real-world systems; however, the accuracy is not great. Many methods have been proposed to improve detection performance such as SSD [25], YOLO-v2 [26], and YOLO-v3 [30], and so on.

12.3 Object detection methods

Numerous object detection methods have been introduced over the past decades. In this section, we present some approaches based on deep CNNs. Fig. 12.3 shows the historical progress of object detection methods.

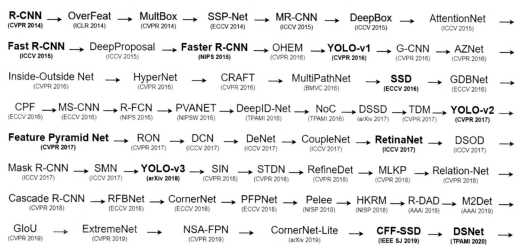

FIG. 12.3 Historical progress of CNN-based object detection methods.

12.3.1 R-CNN

Girshick et al. [20] proposed Regions with CNN features (R-CNN) in 2014, which is known as the first CNN-based detection model. A region proposal-based approach, R-CNN accomplishes object detection in two stages, as shown in Fig. 12.4. In the first stage, the selective search method [21] is employed to generate about 2000 region proposals that may contain objects from the input image. In the second stage, the region proposals are warped into a fixed size, and then they are individually passed through a CNN for feature extraction. Finally, the extracted features of each region are sent to fully connected layers to classify the object and refine the boundary box. R-CNN employs a class-specific linear support vector machine (SVM) for object classification. If the model needs to predict n classes, n SVM classifiers are used.

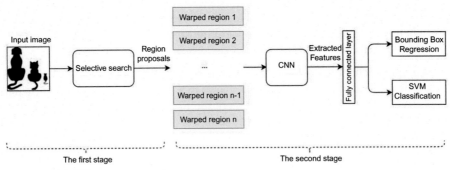

FIG. 12.4   Schematic diagram of R-CNN.

Although R-CNN is superior to previous methods [41, 42] in terms of performance, this method still consumes a huge amount of time in training and inference because of classifying individually a large number of region proposals per image.

12.3.2 Fast R-CNN

Girshick et al. [22] proposed Fast Regions with CNN features (Fast R-CNN) in 2015, which is an improved version of R-CNN, as shown in Fig. 12.5.

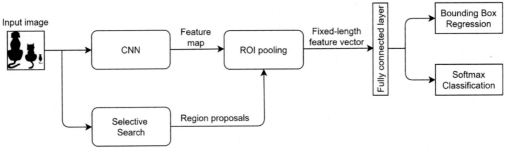

FIG. 12.5   The architecture of Fast R-CNN.

Based on R-CNN, Fast R-CNN incorporates three major changes to improve object detection performance.

- Using a CNN to extract features from the entire image instead of each image patch from scratch, which helps to improve the processing time significantly.
- Using a selective search method to generate RoIs from the input image, which are combined with the corresponding feature maps and sent to an ROI pooling layer to generate fixed-length feature vectors, as shown in Fig. 12.6. These resulting feature vectors are fed into the fully connected layers for object classification and localization.
- Replacing SVM with softmax function for classification.

In Fast R-CNN, RoI pooling is applied to warp the various sizes of ROIs into a fixed size. For example, transforming an 8×8 feature map into a fixed-length feature with a size of 2×2, as shown in Fig. 12.7.

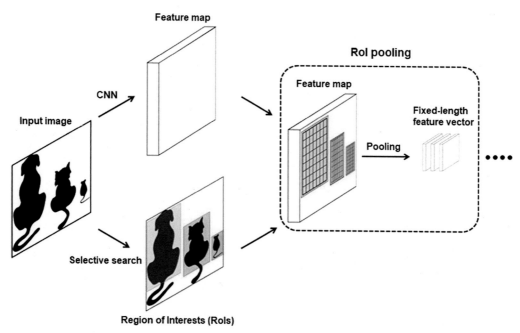

FIG. 12.6 RoI pooling in Fast R-CNN.

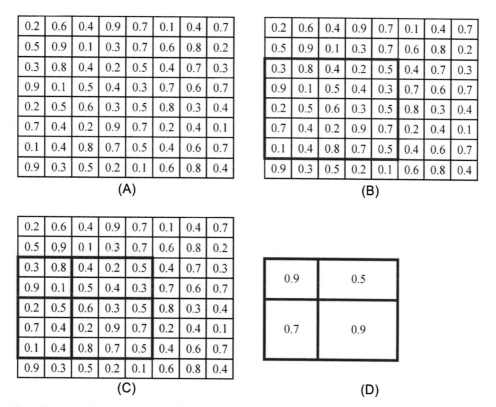

FIG. 12.7 RoI Pooling: (A) an 8×8 feature map, (B) combining the RoI (*red*) with the feature map, (C) splitting RoI into the 2×2 target where the size of four sections can be equal or different, (D) taking a max over each section, resulting in a fixed-length feature map with a size of 2×2.

12.3.3 Faster R-CNN

Ren, et al. [23] proposed Faster Regions with CNN features (Faster R-CNN) in 2015, which is an improved version of Fast R-CNN. Faster R-CNN adopts similar design as the Fast R-CNN except it replaces the selective search method with a region proposal network (RPN) for generating region proposals on the feature map, as shown in Fig. 12.8.

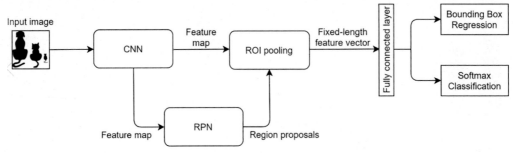

FIG. 12.8   Flowchart of Faster R-CNN.

Faster R-CNN outperforms R-CNN and fast R-CNN in terms of both speed and accuracy. This method has two major changes compared to the previous version:

- Region Proposal Network (RPN): Faster R-CNN employed a CNN, namely ZF [43] or VGG [44] to produce region proposals instead of the selective search method used in R-CNN and Fast R-CNN. As shown in Fig. 12.9, at each sliding-window location, RPN simultaneously predicts multiple region proposals, and it outputs 4k coordinates, and 2k scores that estimate probability of object or not object for each proposal, where k is the number of anchors.

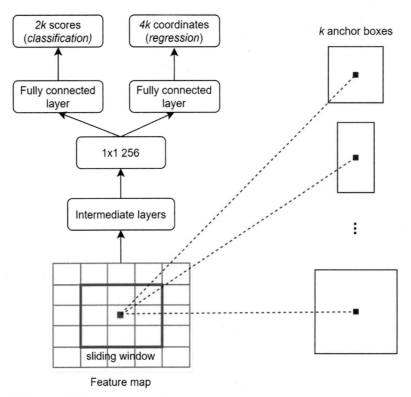

FIG. 12.9   Region Proposal Network (RPN).

- Anchor: Anchor is a reference box, which is related to a scale and aspect ratio. Faster R-CNN used 9 anchor boxes (3 scales and 3 aspect ratios) at each sliding position. Those anchors are pre-selected to cover the various size of real-life objects.

Supplementary explanation

RPN does not output bounding box (x, y, w, h) directly; it employs the parameterizations of the four coordinates as follows:

$$t_x = \frac{x - x_a}{w_a}, t_y = \frac{y - y_a}{h_a}, t_w = \log\left(\frac{w}{w_a}\right), t_h = \log\left(\frac{h}{h_a}\right)$$

$$t_x^* = \frac{x^* - x_a}{w_a}, t_y^* = \frac{y^* - y_a}{h_a}, t_w^* = \log\left(\frac{w^*}{w_a}\right), t_h^* = \log\left(\frac{h^*}{h_a}\right)$$

where x and y represent two coordinates of the predicted box center, w and h represent its width and height, respectively. (x_a, y_a, w_a, h_a) are for the Anchors box, and (x^*, y^*, w^*, h^*) are for the ground-truth box.

12.3.4 YOLO-v1

Redmon et al. [24] proposed You Only Look Once (YOLO or YOLO-v1) in 2016. It is known as the first regression/classification-based object detection method, as shown in Fig. 12.10. Different from the region proposal-based approach, YOLO directly inferences object classification and object localization from the entire input image in one evaluation through a single CNN. This strategy helps YOLO achieve a running time of 45 frames per second, making it a true real-time detector.

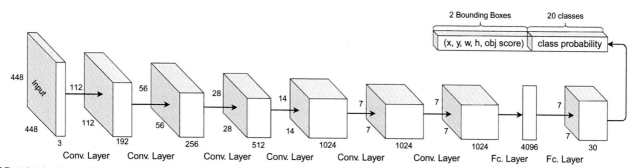

FIG. 12.10 The architecture of YOLO.

For prediction, as shown in Fig. 12.11, the YOLO model divides the input image into an $S \times S$ grid. If a grid cell contains the center of an object, it is accountable for detecting that object. There are B bounding boxes, and confidence scores of those boxes are predicted by each grid cell. Each predicted bounding box contains five predictions including (x, y, w, h) and confidence score, where (x, y) represent the coordinates of the center of the box relative to the bounds of the grid cell, (w, h) represent the width and height which are predicted relative to the whole image, and confidence score reflects whether there is an object. The greater the box confidence score is, the greater the probability that the bounding box predicted by the grid cell contains an object. The class probabilities are based on the number of predicted classes. If there are 20 classes, 20 object class probabilities are output.

The output size of YOLO model is calculated as follows:

$$\text{Output size} = S \times S \times (B \times 5 + C).$$

S: width and height of the grid.

B: number of bounding boxes predicted by each grid cell.

5: each predicted bounding box in the grid cell contains five predictions: (x, y, w, h), and box confidence score.

C: number of conditional class probabilities; if there are 20 categories, 20 class probabilities are output.

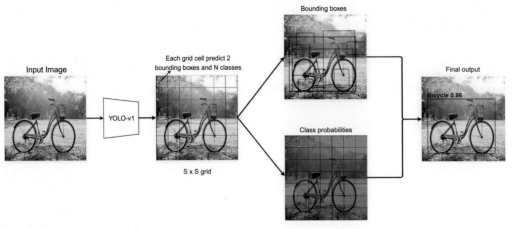

FIG. 12.11 The prediction process of YOLO.

For example, the settings of YOLO on the PASCAL VOC dataset are S=7, B=2, and C=20. The output size of the YOLO model is a $7 \times 7 \times (2 \times 5 + 20) = 7 \times 7 \times 30$ tensor.

Supplementary explanation

YOLO predicts two bounding boxes on each grid cell, so a total of $7 \times 7 \times 2 = 98$ object boxes are predicted. However, 98 bounding boxes are not the final prediction. To keep the best bounding boxes for the final answer, YOLO removes redundant bounding boxes by using the Non-Maximum Suppression (NMS) method, as shown in Fig. 12.12.

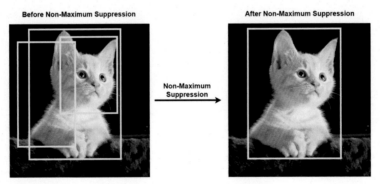

FIG. 12.12 Selecting the best bounding box using Non-Maximum Suppression (NMS).

The NMS method is as follows:

1. If the confidence score is too small, it means that there is no object in the predicted bounding box, so first remove the bounding boxes with a confidence score less than the score threshold.
2. Sort the remaining bounding boxes according to the confidence score from biggest to smallest and put them into a "list object boxes," denoted as L.
3. Take out a bounding box with the highest confidence score from L and put it in a "list of filtered object boxes," denoted as F.
4. Calculate separately the intersection over union (IoU) of the bounding box just put in F with all bounding boxes in L. If the calculated IoU value is greater than the IoU threshold, remove that bounding box from L.
5. Repeat steps 3–4 until there is no bounding box in L.

12.3.5 SSD

Wei Liu et al. [25] proposed Single Shot MultiBox Detector (SSD) in 2016. It is a real-time CNN-based object detection model. The running time prediction of SSD is similar to that of YOLO, but the accuracy is much greater because of differences in architecture design, as shown in Fig. 12.13.

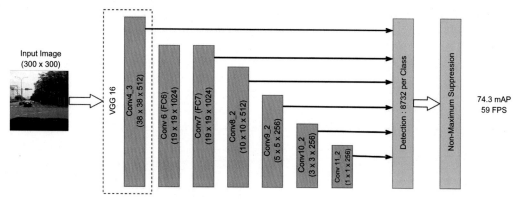

FIG. 12.13 The architecture of Single Shot MultiBox Detector (SSD).

To produce detection, SSD has two main features.

- SSD uses feature maps from six different levels of convolutional layers to make multiscale predictions. In an SSD model, the shallower layers are used for detecting small objects, while the deeper layers are responsible for big object detection. The concept of the prediction in the feature maps outputted by six convolutional layers is similar to that of YOLO. Each grid cell in the feature map predicts multiple bounding boxes, and 8732 bounding boxes per class were predicted by SSD. The number of bounding boxes predicted by the six convolutional layers is calculated as:

The size of feature maps for prediction

$$
\begin{aligned}
\text{Conv4\_3} \;:\; & 38 \times 38 \times 4 = 5{,}776 \\
\text{Conv7} \;\;\;:\; & 19 \times 19 \times 6 = 2{,}166 \\
\text{Conv8\_2} \;:\; & 10 \times 10 \times 6 = 600 \\
\text{Conv9\_2} \;:\; & 5 \times 5 \times 6 = 150 \\
\text{Conv10\_2} :\; & 3 \times 3 \times 4 = 36 \\
\text{Conv11\_3} :\; & 1 \times 1 \times 4 = 4
\end{aligned}
$$

The number of Bounding boxes predicted by each grid in the feature maps

Total bounding box: $5776 + 2166 + 600 + 150 + 36 + 4 = 8732$

- For each predicted bounding box, N class scores and four offsets relative to the default box shape are computed. The default boxes were pre-selected manually to cover a wide range of object sizes in real life. To cope with object prediction at different scales, SSD uses different layers with different sets of default boxes, in which each feature map layer shares the same set of default boxes centered at the corresponding grid.

12.3.6 YOLO-v2

Redmon et al. [26] proposed YOLO-v2, also known as YOLO9000, in 2017. It is an improved version of YOLO that is faster and stronger than both YOLO and SSD. YOLO-v2 incorporates the following seven important changes.

1. Batch normalization: Batch normalization is added in convolutional layers, which helps to avoid overfitting problems during the training process and improves the performance about 2% mean Average Precision (mAP).

2. High-resolution classifier: In YOLO-v2, the classifier is first trained with an input resolution of 224 x 224, then returns with a resolution of 480 x 480 utilizing much fewer epochs. This helps to increase the accuracy of the model by 4% mAP.

3. Convolutional with anchor boxes: YOLO-v2 replaces the prediction of arbitrary bounding boxes used in YOLO with predicting offsets of each given anchor box. Although the accuracy is reduced a little bit when using anchor boxes, it increases the chances for the detection of all ground-truth objects.

4. Dimension clusters: To find top-K boundary boxes that are best suitable for the training dataset, instead of manually selecting anchors, the K-means clustering method is employed to locate the centers of the top-K clusters. For the best design, five anchors were selected for training the YOLO-v2 model.

5. Direct location prediction: Because YOLO does not apply a constraint on location prediction, this makes the network unstable in training, resulting in the predicted bounding boxes being far from the original grid location. Using the sigmoid function to constrain the output prediction of the model to fall in the range between 0 and 1 makes YOLO-v2 more stable in the training process. In the output feature map, YOLO-v2 predicts five bounding boxes at each grid cell; each bounding box has five coordinates (t_x, t_y, t_w, t_h, and t_0). The corresponding prediction is defined as:

$$b_x = \sigma(t_x) + c_x$$
$$b_y = \sigma(t_y) + c_y$$
$$b_w = p_w e^{t_w}$$
$$b_h = p_h e^{t_h}$$

Pr(object)*IOU(predicted box, ground truth box) $= \sigma(t_0)$

b_x, b_y, b_w, b_h: coordinates of the center, width, and height of the predicted box, respectively

c_x, c_y: top left corner of the image

σ: sigmoid activation function

p_w, p_h: width and height of the anchor box, respectively

$\sigma(t_0)$: confidence score of the predicted box

6. Fine-grained features: Unlike SSD, which was run at various feature maps to obtain a range of resolutions, YOLO-v2 concatenates the higher resolution features with the low resolution features for predicting detections. To accomplish the objective, a passthrough layer is employed, which takes a $26 \times 26 \times 512$ feature map as the input and outputs a $13 \times 13 \times 2048$ feature map. Then, the generated features map is concatenated with the original $13 \times 13 \times 1024$ feature map to form a $13 \times 13 \times 3072$ feature map. Finally, convolution filters are applied on the new $13 \times 13 \times 3072$ layer to make predictions, as shown in Fig. 12.14. Following this method, the accuracy of the model improved by about 1% mAP.

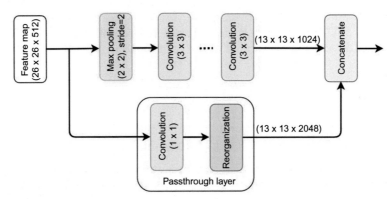

FIG. 12.14   Using fine-grained features of YOLO-v2.

Supplementary explanation

Reorganization layer: Unlike max pooling layer, which reduces the information of the input, the reorganization layer in Fig. 12.14 retains the original information of the input after the operation, as shown in Fig. 12.15.

Reorganization

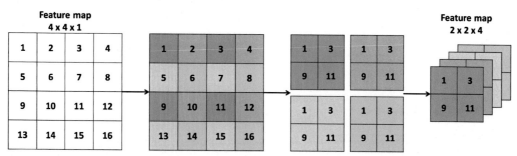

FIG. 12.15 Operation of the reorganization layer.

7. Multi-scale training: During training of the YOLO-v2 model, the size of the input image is changed every 10 batches. By taking multiples of 32 such as 320, 352, 384, 416, 448, 480, 512, 544, 576, and 608, YOLO-v2 uses the minimum input image size of 320 x 320 and maximum size of 608 x 608. This training strategy helps YOLO-v2 to overcome the problems of detecting objects at multiple scales while still operating at real-time speeds.

12.3.7 Feature pyramid networks

Lin et al. [27] introduced Feature Pyramid Networks (FPNs) in 2017. FPNs adopt the feature pyramid architecture with lateral connections for detecting object at different scales, especially small objects. Using the CNN for feature extraction, the deeper layers have the characteristics of low resolution and rich semantic features, while the shallow layers contain features with high resolution and weak semantic information. This leads to the poor performance of object detectors that use the pyramidal feature hierarchy of CNN for detecting objects at multiple scales. To make a feature pyramid that contains rich semantics at all levels, in FPN, the higher-resolution layers are constructed from a semantic-rich layer through a top-down pathway. Because the reconstructed layers have strong semantic features but lack the location information of the objects, lateral connections between these layers and corresponding feature maps are added for better prediction of object location, as shown in Fig. 12.16.

There are four main network architectures to improve the performance of object detection [27]: pyramid of image, single feature map, pyramid of feature maps, and feature pyramid network, as shown in Fig. 12.17. The following summarizes the main characteristics of each architecture.

- Fig. 12.17A shows the network architecture of the pyramid of image network. In this network architecture, features for prediction are independently computed on each of the image scales. Although this method is effective, the running time is very slow.
- Fig. 12.17B shows the network architecture of a single feature map, which uses a single-scale feature for prediction. This approach can achieve real-time prediction speed, but the accuracy is not satisfied because of struggling with small objects. YOLO [24] is a typical representative of this architecture.
- Fig. 12.17C shows the network architecture of pyramid of feature maps or pyramidal feature hierarchy. This architecture is used by SSD [25], which employs feature maps at different scales to make predictions. The advantage of this method is that no additional calculations are required. However, using shallower layers that lack semantic information to predict small objects is the reason for poor performance.
- Fig. 12.17D shows the architecture of an FPN. Similar to the pyramidal feature hierarchy in Fig. 12.17C, FPN [27] also uses feature maps from lower pyramid levels for small object prediction and feature maps from higher pyramid levels for big object prediction. In FPN, high-level semantic information is spread at all scales of feature maps, and therefore it is more effective than the networks in Fig. 12.17C, whereas the speed is similar.

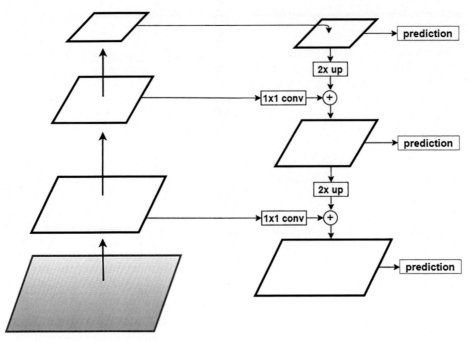

FIG. 12.16   The architecture of a Feature Pyramid Network (FPN).

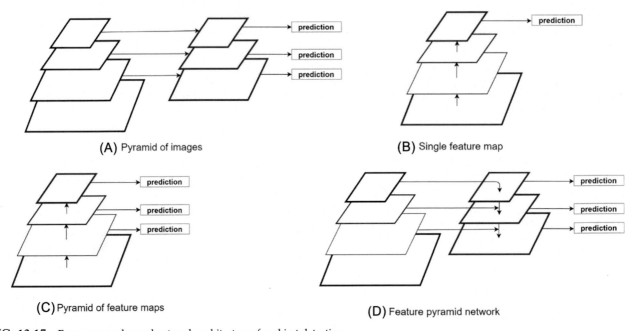

FIG. 12.17   Four commonly used network architectures for object detection.

12.3.8 RetinaNet

Lin et al. [28] proposed Focal Loss for Dense Object Detection (RetinaNet) in 2017. For improving object detection performance, RetinaNet employs ResNet [35] as a backbone, adopts FPN architecture [27] to build a multi-scale feature pyramid for predict object at different scales, and adds two subnets, namely, the classification subnet and box regression subnet, to each pyramid level for object classification and localization, respectively, as shown in Fig. 12.18. The classification subnet and box regression subnet have the same architecture, including four 3 x 3 convolutional layers with ReLU activation, each with 256 filters, followed by a 3 x 3 convolutional layer for prediction. The difference between the two subnets is that the last layer of the classification subnet uses $A * K$ filters, followed by sigmoid

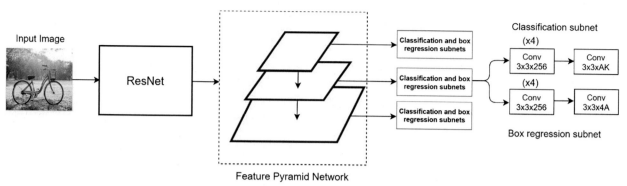

FIG. 12.18 The architecture of RetinaNet.

activation function, while that of the box regression subnet utilizes *4 * A* filters. *A* and *K* represent the number of anchors and object classes, respectively. Noteworthy, focal loss is proposed for training object classification in Retina-Net to solve the class imbalance problem and increase the accuracy of object detection.

Focal loss is an improved version of cross entropy loss ($CE = -\log(p_t)$), which is defined as:

$$FL_{(p_t)} = -\alpha_t (1 - p_t)^\gamma \log(p_t),$$

where α_t is a weighting factor ($\alpha_t \in [0,1]$), γ is a tunable focusing parameter ($\gamma \geq 0$), and p_t is defined as:

$$p_t = \begin{cases} p & if \ y = 1 \\ 1 - p & otherwise, \end{cases}$$

here, y is the ground-truth class ($y \in \{\pm 1\}$), p is the predicted probability for the class with $y = 1$ of the model.

If $\gamma = 0$, $FL_{(p_t)}$ is equal to CE, if setting $\gamma > 0$, the network can focus more on misclassified examples instead of well-classified examples during training process. For example:

- The network model outputs a high probability value, such as $p_t = 0.8$, which means the prediction result is close to the ground truth (well-classified example).

  When $\gamma = 0$, FL = 0.32192
  When $\gamma = 2$, FL = 0.01287

- The network model outputs a small probability value, such as $p_t = 0.2$, which means the prediction result is very different from the ground truth (misclassified example).

  When $\gamma = 0$, FL = 2.32192
  When $\gamma = 2$, FL = 1.48603

As shown, if $\gamma > 0$, the loss for well-classified examples is dow-weighted ($\gamma = 2$, $p_t = 0.8$, FL = 0.01287), and the model focuses on misclassified example ($\gamma = 2$, $p_t = 0.2$, FL = 1.48603). In the experiments, RetinaNet works best with $\gamma = 2$.

12.3.9 YOLO-v3

Redmon proposed YOLO-v3 [30], which is an optimized version of YOLO-v2, in 2018. YOLO-v3 obtains impressive detection performance through three main improvements: (1) a change of backbone network for better feature extraction, (2) increasing the number of anchor boxes for more accurate predictions, and (3) applying the concept of FPN for multiscale object detection, as shown in Fig. 12.19.

1. Darknet-53 backbone: The first step of all object detection methods is to train a backbone network for performing feature extraction. For example, the backbone of SSD [25] is VGG-16, the backbone of YOLO-v2 [26] is DarkNet-19, the backbone of RetinaNet [28] is ResNet-101 [35], and the backbone of YOLO-v3 is Darknet-53. The Darknet-53 network is built on Darknet-19 architecture by adding convolutional layers with some shortcut connections. It consists of 53 convolutional layers, as shown in Fig. 12.20. Darknet-53 has proven more efficient than DarkNet-19, ResNet-101, and ResNet-152 in terms of accuracy and billion floating point operations per second [30].

Input shape: (416, 416, 3)
Classes: 80
Output shape: ($grid_h$, $grid_w$, anchor, (5 + Classes))

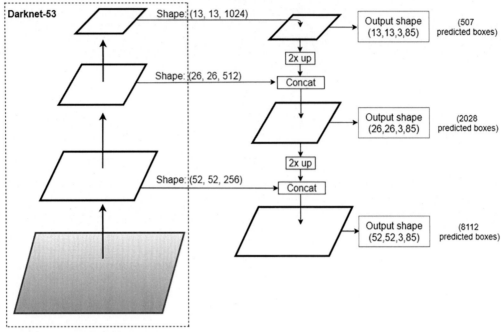

FIG. 12.19 The architecture of YOLO-v3.

Type	Number of kernels	Kernel size	Stride	Output
Convolution	32	3 x 3		224 x 224
Maxpooling		2 x 2	2	112 x 112
Convolution	64	3 x 3		112 x 112
Maxpooling		2 x 2		56 x 56
Convolution	128	3 x 3		56 x 56
Convolution	64	1 x 1		56 x 56
Convolution	128	3 x 3		56 x 56
Maxpooling		2 x 2	2	28 x 28
Convolution	256	3 x 3		28 x 28
Convolution	128	1 x 1		28 x 28
Convolution	256	3 x 3		28 x 28
Maxpooling		2 x 2	2	14 x 14
Convolution	512	3 x 3		14 x 14
Convolution	256	1 x 1		14 x 14
Convolution	512	3 x 3		14 x 14
Convolution	256	1 x 1		14 x 14
Convolution	512	3 x 3		14 x 14
Maxpooling		2 x 2	2	7 x 7
Convolution	1024	3 x 3		7 x 7
Convolution	512	1 x 1		7 x 7
Convolution	1024	3 x 3		7 x 7
Convolution	512	1 x 1		7 x 7
Convolution	1024	3 x 3		7 x 7
Convolution	1000	1 x 1		7 x 7
Avgpooling		Global		1000
Softmax				

(a) Darknet-19

Type	Number of kernels	Kernel size	Stride	Output	
Convolution	32	3 x 3		256 x 256	
Convolution	64	3 x 3	2	128 x 128	
Convolution	32	1 x 1			
Convolution	64	3 x 3			x1
Residual				128 x 128	
Convolution	128	3 x 3	2	64 x 64	
Convolution	64	1 x 1			
Convolution	128	3 x 3			x2
Residual				64 x 64	
Convolution	256	3 x 3	2	32 x 32	
Convolution	128	1 x 1			
Convolution	256	3 x 3			x8
Residual				32 x 32	
Convolution	512	3 x 3	2	16 x 16	
Convolution	256	1 x 1			
Convolution	512	3 x 3			x8
Residual				16 x 16	
Convolution	1024	3 x 3	2	8 x 8	
Convolution	512	1 x 1			
Convolution	1024	3 x 3			x4
Residual				8 x 8	
Avgpooling					
Conneted					
Softmax					

(b) Darknet-53

FIG. 12.20 The architecture of Darknet-19 and Darknet-53.

2. Anchor boxes and multiscale prediction: YOLO-v2 used five anchor boxes to predict the location of objects, while YOLO-v3 increases the number of anchor boxes from five to nine. These nine anchor boxes with different scales are assigned to three different output feature maps, as shown in Table 12.1.

TABLE 12.1 Anchor boxes used in YOLO-v3.

Size of output feature maps	Size of anchor boxes
52×52	$10 \times 13, 16 \times 30, 33 \times 23$
26×26	$30 \times 61, 62 \times 45, 59 \times 119$
13×13	$116 \times 90, 156 \times 198, 373 \times 326$

Three outputs of YOLO-v3 are 52×52, 26×26, and 13×13 grid cells, corresponding to the output feature maps of 52×52, 26×26, and 13×13. The smaller the output feature map is, the larger the receptive field of the corresponding grid cell is. As shown in Fig. 12.21, there are 4×4 and 2×2 feature maps. Each grid cell of the 4×4 feature map has a smaller field of view than that of a 2×2 feature map, so using a small anchor box of (a) is better than using big anchor boxes of (c). In contrast, each grid cell of the 2×2 feature map has a larger field of view, so using the big anchor box of (d) is better than using the small anchor boxes of (b).

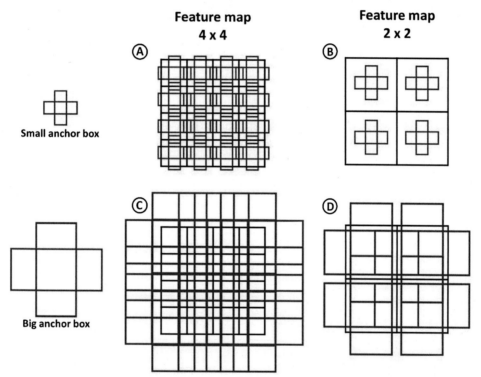

FIG. 12.21 Configuration of anchor box.

Since three different anchor boxes are assigned to each output feature map, and each grid cell predicts three bounding boxes, the shape of each output feature map of YOLO-v3 can be expressed as $grid_h \times grid_w \times 3 \times (4+1+c)$ for the 4 bounding box offsets, 1 objectness prediction, and c class prediction. For example, using YOLO-v3 to detect objects on the COCO dataset of 80 classes ($c = 80$). The output shapes of three prediction layers are (13,13,3,85), (26,26,3,85), and (52,52,3,85), and the corresponding number of predicted boxes are 507, 2028 and 8112, respectively, as shown in Fig. 12.19.

To train YOLO-v3, the objective loss function consists of three parts: offset loss (xy_{loss} and wh_{loss}), bounding box confidence loss, and classification loss, which are defined as follows:

$$xy_{loss} = \lambda_{coord} \sum_{i=0}^{S^2} \sum_{j=0}^{B} \mathbb{1}_{i,j}^{obj} \left[\left(t_{x_i} - \hat{t}_{x_i}\right)^2 + \left(t_{y_i} - \hat{t}_{y_i}\right)^2 \right]$$

$$wh_{loss} = \lambda_{coord} \sum_{i=0}^{S^2} \sum_{j=0}^{B} \mathbb{1}_{i,j}^{obj} \left[\left(t_{w_i} - \hat{t}_{w_i}\right)^2 + \left(t_{h_i} - \hat{t}_{h_i}\right)^2 \right]$$

$$confidence_{loss} = \sum_{i=0}^{S^2} \sum_{j=0}^{B} \mathbb{1}_{i,j}^{obj} BCE\left(C_i, \hat{C}_i\right) + \sum_{i=0}^{S^2} \sum_{j=0}^{B} \mathbb{1}_{i,j}^{noobj} BCE\left(C_i, \hat{C}_i\right)$$

$$class_{loss} = \sum_{i=0}^{S^2} \sum_{j=0}^{B} \mathbb{1}_{i,j}^{obj} \sum_{c \in classes} BCE\left(p_i(c), \hat{p}_i(c)\right)$$

t_x, t_y, t_w, t_h: four coordinates for each ground truth bounding box

$\hat{t}_x, \hat{t}_y, \hat{t}_w, \hat{t}_h$: four coordinates for each predicted bounding box

C_i: expected box confidence score of box in cell i

$\hat{C}_i$: predicted box confidence score of box in cell i

$p_i(c)$: expected class probabilities for class c in cell i

$\hat{p}_i(c)$: predicted class probabilities for class c in cell i

S: width and height of the grid; there are three sizes of grid in YOLO-v3 including 13×13, 26×26, and 52×52

B: number of bounding boxes predicted by each grid cell

$\mathbb{1}_{i,j}^{obj} = 1$ if the jth bounding box in cell i is responsible for detecting the object, otherwise it is 0

$\mathbb{1}_{i,j}^{noobj} = 1$ if there is no object in cell i

BCE: binary cross-entropy

12.3.10 CFF-SSD

Le et al. [31] proposed Cross-Resolution Feature Fusion Single Shot MultiBox Detector (CFF-SSD) in 2019. CFF-SSD improves small object detection performance, especially detection of small human hands in intelligent homecare systems. There are two versions of CFF-SSD: 1CFF-SSD for fast detection and 2CFF-SSD for more accurate detection, as shown in Fig. 12.22.

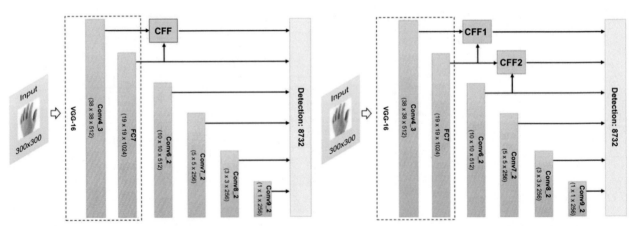

FIG. 12.22   The architecture of CFF-SSD [31].

CFF-SSD achieves impressive small object detection performance because of applying the proposed cross-resolution feature fusion (CFF) approach to the base detector SSD [25] for enriching semantic and contextual information in shallow convolutional layers. The CFF architecture includes two main modules: (1) Narrow Atrous Spatial Pyramid Pooling (N-ASPP), and (2) Richer Semantic Information Generation (RSIG) module, as shown in Fig. 12.23.

1. Narrow Atrous Spatial Pyramid Pooling(N-ASPP) module

The N-ASPP module uses a feature map at a shallow layer as the input and extracts multi-scale contextual information through four parallel atrous convolutions with different rates. In 1CFF-SSD architecture, one CFF is adopted, which uses N-ASPP with four parallel atrous convolutions including a 1×1 convolution and three 3×3 atrous convolutions with rates of 1, 2, and 4, respectively. In 2CFF-SSD architecture, one more CFF is added to the 1CFF-SSD architecture. Unlike the first CFF, the second CFF adopts an N-ASSP module using atrous convolutions with bigger

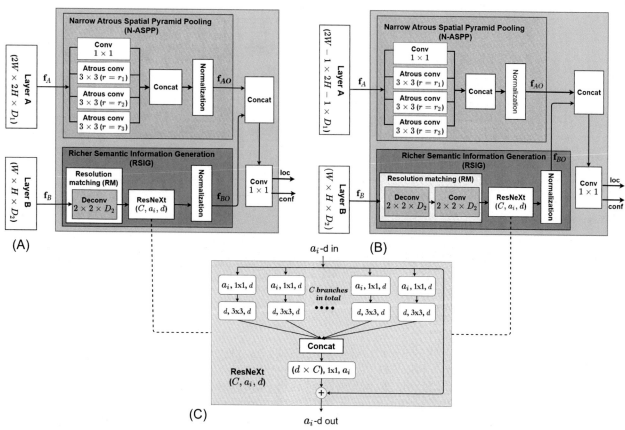

FIG. 12.23 Cross-Resolution Feature Fusion (CFF) [31].

rates to enlarge the field of view of filters for combining larger context. Specifically, three rates of three 3×3 atrous convolutions in the N-ASSP module of the second CFF are set to 2, 4, and 8, respectively. Although the output of the N-ASSP module contains rich contextual information, it lacks semantic information. Therefore, it is combined with the output of the Richer Semantic Information Generation (RSIG) module to supplement semantic information before using for prediction.

2. Richer Semantic Information Generation (RSIG) module

The RSIG module is responsible for providing rich semantic information from a deeper layer to a shallower layer and consists of two modules: a resolution matching (RM) submodule and the ResNeXt submodule. First, the RM module enlarges the feature maps of the deeper layer to the same size as the output of the N-ASPP module. If the input size of the RM module and the output size of the N-ASPP module are $(W \times H \times D)$ and $(2W \times 2H \times D)$, respectively, the RM module only contains one deconvolutional layer, as shown in Fig. 12.23A. If the input size of the RM module and the output size of the N-ASPP module are $(W \times H \times D)$ and $(2W-1 \times 2H-1 \times D)$, respectively, two deconvolutional layers are adopted in the RM module, as shown in Fig. 12.23B. Next, the ResNeXt module uses the enlarged feature maps from the RM module as the input and exploits richer semantic information through multiple branches of convolutional layers. Finally, the resulting feature of ResNeXt module is passed through a normalization layer for producing a feature map at the output of RSIG module. The resulting feature map of the RSIG module is concatenated with that of the N-ASPP module for improving the performance of small object detection.

Supplementary explanation

Atrous convolution: Proposed in DeepLab [45], atrous convolution adjusts the field of view of the filter by inserting zeros between adjacent filter elements, as shown in Fig. 12.24. The number of zeros can be controlled by setting a hyperparameter "rate." Atrous convolution with rate r indicates that $(r-1)$ zeros are padded, and the bigger the "rate" value used, the larger the

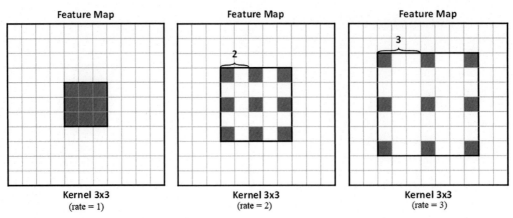

FIG. 12.24   Atrous convolution with different rate values.

field of view of the filter becomes. Using atrous convolution with different atrous rates effectively extracts multi-scale context information from a feature map without any additional computation.

12.3.11 DSNet

Huang et al. [32] introduced the Dual-Subnet Network (DSNet) in 2021. DSNet is designed based on multi-task learning to improve the performance of object detection in inclement weather conditions, namely, foggy weather. To accomplish this objective, DSNet employs RetinaNet [28] as the base detector, denoted as a detection subnetwork, and adds a feature recovery (FR) module to this base architecture to build a restoration subnetwork for enhancing the visibility of degraded images, as shown in Fig. 12.25.

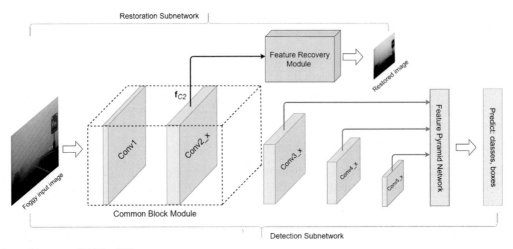

FIG. 12.25   The architecture of DSNet [32].

1. The detection subnetwork

In the detection subnetwork, ResNet-50 [35] with 500 pixel input image scale is first employed as the backbone for performing feature extraction of the entire input image. Then, a feature pyramid network [27] is constructed at the top of the backbone for multi-scale object detection. Finally, two task-specific subnets: class subnet and box subnet, are attached at each feature pyramid level to perform object classification and localization, respectively, as shown in Fig. 12.26.

Both class subnet and box subnet are fully convolutional networks and composed of five convolution layers with filter size of 3×3, in which each of the first four convolution layers is followed by rectified linear unit (ReLU) activation. The difference between the class subnet and box subnet is in the last layer. While the sigmoid activation function is

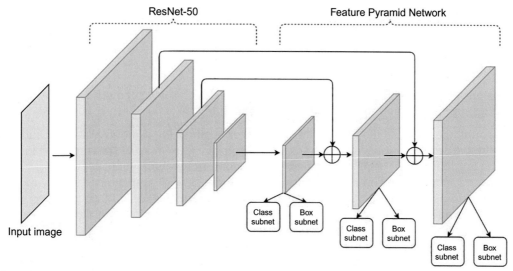

FIG. 12.26 The architecture of a detection subnetwork.

attached to the last layer with c*A filters of the class subnet for binary predictions, the box subnet uses the last layer with 4*A filters for linear prediction, where c and A represent the number of object classes and anchor boxes, respectively. Please refer to Section 12.3.9 for more detail about the anchor box.

2. The restoration subnetwork

The restoration subnetwork is designed based on the transformed atmospheric scattering model and is composed of two main modules: a common block module and a feature recovery module, as shown in Fig. 12.27.

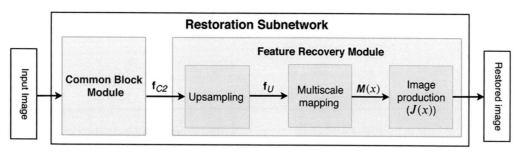

FIG. 12.27 Flowchart of restoration subnetwork [32].

(a) Transformed atmospheric scattering model.

The atmospheric scattering model is expressed as:

$$I(x) = J(x)t(x) + \propto (1 - t(x)),$$

where $I(x)$ and $J(x)$ represent the hazy image and haze-free image, respectively, $\propto$ is the global atmospheric light, and $t(x)$ is the medium transmission defined as:

$$t(x) = e^{-\beta d(x)}$$

Here, β and $d(x)$ are the scattering coefficient of the atmosphere and the scene depth, respectively.

From the atmospheric scattering model, the haze-free image is calculated as:

$$J(x) = M(x)I(x) - M(x) + 1$$

where $M(x) = \dfrac{\frac{1}{t(x)}(I(x) - \propto) + (\propto - 1)}{I(x) - 1}$

The estimation of $M(x)$ and $J(x)$ is presented below.

(b) Common block module

The detection subnetwork shares a common block module containing the first 10 convolution layers with the restoration subnetwork to guarantee clean features produced by this module can be computed for detecting objects.

(c) Feature recovery (FR) module

To enhance visibility in the image effectively, the FR is designed with three submodules: upsampling submodule, multiscale mapping submodule, and image production submodule.

- Upsampling submodule: Because the restoration subnetwork requires the same size of input and output, the upsampling submodule is used to increase the size of the output feature map of the common block module to the same size as the input image by using the bilinear interpolation technique.
- Multiscale mapping submodule: This module adopts four parallel layers, namely, one convolutional layer with a 1×1 filter, one convolutional layer with a 3×3 filter, one convolutional layer with a 5×5 filter, and one convolutional layer with a 7×7 filter for multi-scale feature extraction. The extracted features are sent to another convolutional layer with a 3×3 filter for computing M(x).
- Image production submodule: This module takes M(x) as the input and uses element-wise multiplication, element-wise subtraction, and element-wise addition operations for computing a haze-free image J(x).

DSNet is trained in an end-to-end fashion for jointly learning three tasks including visibility enhancement, object classification, and object localization by using the objective function:

$$L_{DSNet} = L_{cls} + L_{loc} + L_v$$

L_{cls}: focal loss [28] is adopted for object classification

   L_{loc}: smooth L_1 loss is used for box regression [22]

   L_v: loss function for visibility enhancement; this is mean square error (MSE) loss used for a ground-truth image Y and an estimated image $\hat{Y}$

$$M(x) = \frac{1}{m} \sum_{i}^{m} \left(Y_i - \hat{Y}_i \right)^2$$

where m is the batch size.

12.4 Experiment: Implementation of YOLO-v3

This section introduces how to build the YOLO-v3 model for object detection. To conduct experiments, we used the PASCAL VOC 2007 dataset for training and testing the model.

12.4.1 Load YOLO-v3 project

Because the example program of YOLO-v3 model in this chapter is more complicated than the previous examples in Chapters 1–9, we use PyCharm IDE as a compiler to write and train the model. The process of loading the YOLO-v3 project is as follows.

1. Download the YOLO-v3 project: please go to https://github.com/taipeitechmmslab/MMSLAB-DL/tree/master to download the source code, as shown in Fig. 12.28.

FIG. 12.28   YOLO-v3 project (Lab12) on GitHub.

2. Use PyCharm to open the project, as shown in Fig. 12.29.

FIG. 12.29   Open the project on PyCharm.

3. Select the path to load the project, as shown in Fig. 12.30.

FIG. 12.30   The path of the YOLO-v3 project.

4. Set the compilation environment (Note: if this step already set, it is omitted): Click File → Settings → Project "name of the project" → Project Interpreter → Add → Virtualenv Environment → Existing environment → Select the virtual machine, as shown in Figs. 12.31 and 12.32.

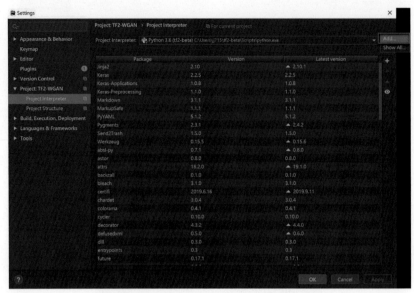

FIG. 12.31   New interpreter.

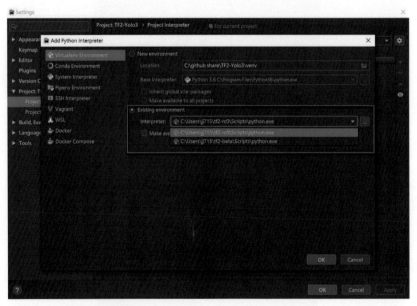

FIG. 12.32   Selecting an existing virtual environment.

12.4.2 Introduction to dataset

In this section, we use the PASCAL Visual Object Classes 2007 (PASCAL VOC 2007) dataset to train and test the YOLO-v3 model. This is a benchmark dataset that has been used to evaluate many object detection methods, such as R-CNN, SSD, YOLO, and so on. The dataset can be loaded through TensorFlow datasets as follows.

```
import tensorflow_datasets as tfds
# Load voc2007 dataset
train_data, info = tfds.load("voc2007", split= tfds.Split.TRAIN, with_info=True)
valid_data = tfds.load("voc2007", split= tfds.Split.VALIDATION)
test_data = tfds.load("voc2007", split= tfds.Split.TRAIN)
# Display information about the dataset
print(info)
```

Result:

```
tfds.core.DatasetInfo(
    name='voc2007',
    version=1.0.0,
    description='This dataset contains the data from the PASCAL Visual Object Classes Challenge
2007, a.k.a. VOC2007, corresponding to the Classification and Detection
competitions.
A total of 9,963 images are included in this dataset, where each image contains
a set of objects, out of 20 different classes, making a total of 24,640
annotated objects.
In the Classification competition, the goal is to predict the set of labels
contained in the image, while in the Detection competition the goal is to
predict the bounding box and label of each individual object.
',
    urls=['http://host.robots.ox.ac.uk/pascal/VOC/voc2007/'],
    features=FeaturesDict({
        'image': Image(shape=(None, None, 3), dtype=tf.uint8),
        'image/filename': Text(shape=(), dtype=tf.string, encoder=None),
        'labels': Sequence(shape=(None,), dtype=tf.int64, feature=ClassLabel(shape=(), dtype=tf.int64, num_classes=20)),
        'labels_no_difficult': Sequence(shape=(None,), dtype=tf.int64, feature=ClassLabel(shape=(), dtype=tf.int64, num_classes
=20)),
        'objects': SequenceDict({'label': ClassLabel(shape=(), dtype=tf.int64, num_classes=20), 'bbox': BBoxFeature(shape=(4,),
dtype=tf.float32), 'pose': ClassLabel(shape=(), dtype=tf.int64, num_classes=5), 'is_truncated': Tensor(shape=(), dtype=tf.boo
l), 'is_difficult': Tensor(shape=(), dtype=tf.bool)})
    },
    total_num_examples=9963,
    splits={
        'test': <tfds.core.SplitInfo num_examples=4952>,
        'train': <tfds.core.SplitInfo num_examples=2501>,
        'validation': <tfds.core.SplitInfo num_examples=2510>
    },
    supervised_keys=None,
    citation='"""
        @misc{pascal-voc-2007,
            author = "Everingham, M. and Van~Gool, L. and Williams, C. K. I. and Winn, J. and Zisserman, A.",
            title = "The {PASCAL} {V}isual {O}bject {C}lasses {C}hallenge 2007 {(VOC2007)} {R}esults",
            howpublished = "http://www.pascal-network.org/challenges/VOC/voc2007/workshop/index.html"}

    """,
    redistribution_info=,
)
```

The PASCAL VOC 2007 dataset consist of three sets, including a training set with 2501 images, validation set with 2510 images, and test set with 4952 images. This dataset has 20 classes, which can be displayed by using the following command.

```
info.features['labels'].names
```

Result: ['aeroplane', 'bicycle', 'bird', 'boat', 'bottle', 'bus', 'car', 'cat', 'chair', 'cow', 'diningtable',
'dog', 'horse', 'motorbike', 'person', 'pottedplant', 'sheep', 'sofa', 'train', 'tvmonitor']

Display dataset: display an image in the dataset and ground-truth bounding boxes of objects

```
classes_list = info.features['labels'].names  # class labels
for dataset in train_data.take(1):
    img = dataset['image']  # Read image
    bboxes = dataset['objects']['bbox']  # Read bounding box
    labels_index = dataset['objects']['label']  # Read label (class)
    img = img.numpy()  # Convert the image to Numpy format
    h, w, _ = img.shape  # Read image shape
    for box, label_index in zip(bboxes, labels_index):
        # box [y1, x1, y2, x2] , (y1, x1) is top left corner, (y2, x2) is bottom right corner of the box.
        # the values of box coordinates [y1, x1, y2, x2] is between 0~1.
```

```
        x1 = tf.cast(box[1]*w, tf.int16).numpy()
        y1 = tf.cast(box[0]*h, tf.int16).numpy()
        x2 = tf.cast(box[3]*w, tf.int16).numpy()
        y2 = tf.cast(box[2]*h, tf.int16).numpy()
        # Draw ground-truth bounding box of object in the image
        cv2.rectangle(img, (x1, y1), (x2, y2), (0, 255, 0), 2)
        # Mark the label in the top left corner of the bounding box
        cv2.putText(img,                    # the image to be drawn
                classes_list[label_index],   # class name to be drawn
                (x1, y1 - 3),               # Where to place the text
                cv2.FONT_HERSHEY_SIMPLEX,    # Font style
                1, (0, 244, 0), 2)          # Font size, font color, font thickness
    plt.imshow(img)
```

Result:

12.4.3 Building YOLO-v3 model

1. Directory and files

Fig. 12.33 shows the YOLO-v3 project, which includes the following files and folders.

FIG. 12.33 The YOLO-v3 project.

- train.py: the source code file for training the YOLO-v3 model
- train-multi-scale.py: the source code file for multi-scale training
- test.py: the source code file for testing the model
- config.py: the source code file for setting model parameters, training parameters, and weight path
- convert.py: the source code file for converting the trained weight of YOLO-v3 into the TensorFlow format
- layers: the folder where the source code files of custom network layers are stored
- model: the folder where the source code file of custom network models is stored
- losses: the folder where the source code file of custom loss function is stored
- utils: the folder where the toolkits are stored, such as dataset.py for loading the dataset, evaluation.py for evaluating the model, and so on
- output_images: the folder for storing output results

Running the YOLO-v3 project requires executing two files: "train.py" (or "train-multi-scale.py") for training the model and "test.py" for testing the trained model. The difference between "train.py" and "train-multi-scale.py" is that "train.py" uses a fixed input size to train the model, whereas "train-multi-scale.py" employs multi-scale training, where the size of the input image is changed by multiples of 32, such as 320, 352, 384, 416, 448, 480, 512, 544, 576, and 608, every 10 epochs of training. The minimum input size is 320×320 and the maximum input size is 608×608. The result of the training method using "train-multi-scale.py" is better than that of using "train.py," but it takes a longer time to train.

2. Implementing the YOLO-v3 model

Fig. 12.34 is the flowchart of the source code for building the YOLO-v3 model.

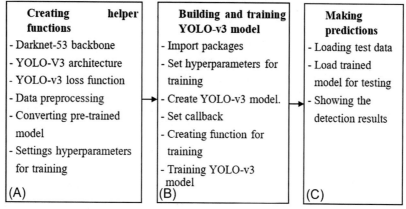

FIG. 12.34 Flowchart of the source code for the YOLO-v3 model.

(a) Creating helper functions

- Darknet-53 backbone: The source code for Darknet-53 is written at "model → darknet.py." YOLO-v3 employed Darknet-53 as the backbone network for feature extraction. Figs. 12.35 and 12.36 show the architecture of the Darknet-53.

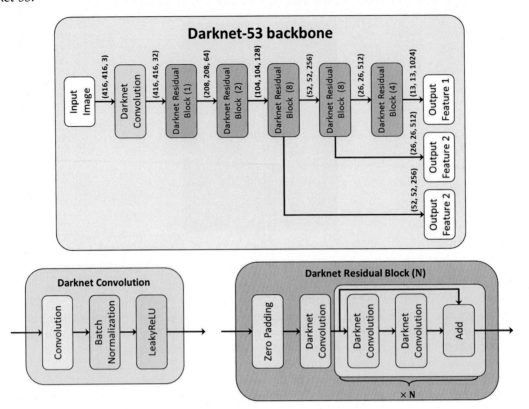

FIG. 12.35   The architecture of Darknet-53 (1).

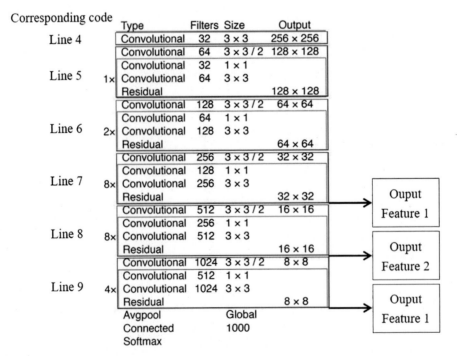

Corresponding code

	Type	Filters	Size	Output
Line 4	Convolutional	32	3 × 3	256 × 256
	Convolutional	64	3 × 3 / 2	128 × 128
	Convolutional	32	1 × 1	
Line 5 1×	Convolutional	64	3 × 3	
	Residual			128 × 128
	Convolutional	128	3 × 3 / 2	64 × 64
	Convolutional	64	1 × 1	
Line 6 2×	Convolutional	128	3 × 3	
	Residual			64 × 64
	Convolutional	256	3 × 3 / 2	32 × 32
	Convolutional	128	1 × 1	
Line 7 8×	Convolutional	256	3 × 3	
	Residual			32 × 32
	Convolutional	512	3 × 3 / 2	16 × 16
	Convolutional	256	1 × 1	
Line 8 8×	Convolutional	512	3 × 3	
	Residual			16 × 16
	Convolutional	1024	3 × 3 / 2	8 × 8
	Convolutional	512	1 × 1	
Line 9 4×	Convolutional	1024	3 × 3	
	Residual			8 × 8
	Avgpool		Global	
	Connected		1000	
	Softmax			

Ouput Feature 1

Ouput Feature 2

Ouput Feature 1

FIG. 12.36   The architecture of Darknet-53 (2).

The code for Darknet-53 is as follows.

Line 1	def darknet_body(name=None):
Line 2	''' Create the Darknet-53 network'''
Line 3	x = inputs = tf.keras.Input([None, None, 3])
Line 4	x = darknetconv2d_bn_leaky(x, 32, (3, 3))
Line 5	x = resblock_body(x, 64, 1)
Line 6	x = resblock_body(x, 128, 2)
Line 7	x = x_26 = resblock_body(x, 256, 8)
Line 8	x = x_43 = resblock_body(x, 512, 8)
Line 9	x = resblock_body(x, 1024, 4)
Line 10	return tf.keras.Model(inputs, (x_26, x_43, x), name=name)

Code path
model→darknet.py

The source codes for Darknet Convolution and Darknet Residual Block are as follows.

```python
import tensorflow as tf
from tensorflow.keras import layers

# Darknet Convolution in Figure 12.35
def darknetconv2d_bn_leaky(x, filters, kernel_size, strides=(1, 1)):
    padding = 'valid' if strides == (2, 2) else 'same'
    x = layers.Conv2D(filters, kernel_size, strides,
                padding=padding,
                use_bias=False,
                kernel_regularizer=tf.keras.regularizers.l2(5e-4))(x)
    x = layers.BatchNormalization()(x)
    x = layers.LeakyReLU(alpha=0.1)(x)
    return x


# the Darknet residual block in Figure 12.35
def resblock_body(x, num_filters, num_blocks):
    x = layers.ZeroPadding2D(((1, 0), (1, 0)))(x)
    x = darknetconv2d_bn_leaky(x, num_filters, (3, 3), strides=(2, 2))
    for i in range(num_blocks):
        y = darknetconv2d_bn_leaky(x, num_filters//2, (1, 1))
```

Code path
model→darknet.py

```python
        y = darknetconv2d_bn_leaky(y, num_filters, (3, 3))
        x = layers.Add()([x, y])
    return x
```

- YOLO-v3 architecture: The source code for the YOLO-v3 model is written at "model→yolo.py." YOLO-v3 uses three different convolutional layers to detect objects at multiple scales. Figs. 12.37 and 12.38 show the architecture of YOLO-v3.

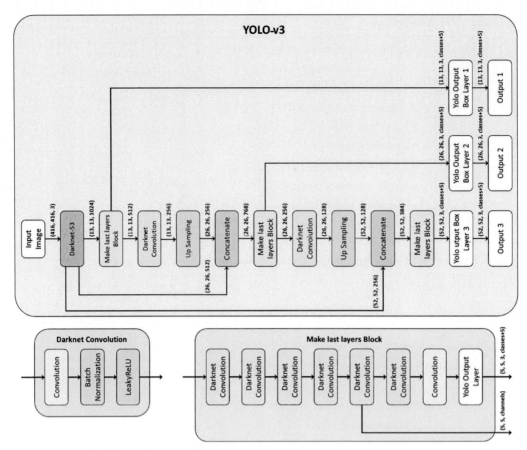

FIG. 12.37   The architecture of YOLO-v3 (1).

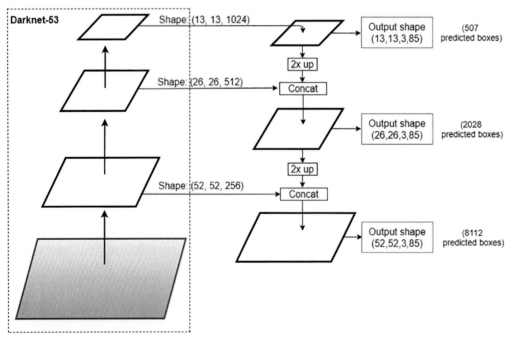

FIG. 12.38   The architecture of YOLO-v3 (2).

The source code of YOLO-v3 model is as follows.

```
def yolov3(input_size, anchors=yolo_anchors, num_classes=80,
           iou_threshold=0.5, score_threshold=0.5, training=False):
    """ Create YOLO_V3 network  (training mode or test mode))"""
    # Using 3 Anchors in each prediction layer
    num_anchors = len(anchors) // 3
    # Declare the Input
    inputs = Input(input_size)
    # Create Darknet-53 network with three output layers
    x_26, x_43, x = darknet_body(name='Yolo_DarkNet')(inputs)
    # layer y1 with output shape: (13, 13, 3, classes+5)
    x, y1 = make_last_layers(x, 512, num_anchors, num_classes)
    x = darknetconv2d_bn_leaky(x, 256, (1, 1))
    # Upsampling (13, 13, 256) -> (26, 26, 256)
    x = layers.UpSampling2D(2)(x)
    # Concat (26, 26, 256) + (26, 26, 512) = (26, 26, 768)
    x = layers.Concatenate()([x, x_43])


    # layer y2 with output shape: (26, 26, 3, classes+5)
    x, y2 = make_last_layers(x, 256, num_anchors, num_classes)
    x = darknetconv2d_bn_leaky(x, 128, (1, 1))
    # Upsampling (26, 26, 128) -> (52, 52, 128)
    x = layers.UpSampling2D(2)(x)
```

Code path
model→yolo.py

```
# Concat (52, 52, 128) + (52, 52, 256) = (52, 52, 384)
    x = layers.Concatenate()([x, x_26])

# layer y3 with output shape: (52, 52, 3, classes+5)
x, y3 = make_last_layers(x, 128, num_anchors, num_classes)
# get input size
h, w, _ = input_size
# Convert the output (tx, ty, tw, th) into (x1, y1, x2, y2)
y1 = YoloOutputBoxLayer(anchors[6:], 1, num_classes, training)(y1)
y2 = YoloOutputBoxLayer(anchors[3:6], 2, num_classes, training)(y2)
y3 = YoloOutputBoxLayer(anchors[0:3], 3, num_classes, training)(y3)
# If it is in training mode, create a training network model
if training:
    return Model(inputs, (y1, y2, y3), name='Yolo-V3')
# using NMS for removing redundance prediction boxes
outputs = NMSLayer(num_classes, iou_threshold, score_threshold)([y1, y2, y3])
return Model(inputs, outputs, name='Yolo-V3')


# "Make last layers Block" in Figure 12.50
def make_last_layers(x, num_filters, num_anchors, num_classes):
    out_filters = num_anchors * (num_classes + 5)
    x = darknetconv2d_bn_leaky(x, num_filters, (1, 1))
    x = darknetconv2d_bn_leaky(x, num_filters * 2, (3, 3))
    x = darknetconv2d_bn_leaky(x, num_filters, (1, 1))
    x = darknetconv2d_bn_leaky(x, num_filters * 2, (3, 3))
    x = darknetconv2d_bn_leaky(x, num_filters, (1, 1))
    y = darknetconv2d_bn_leaky(x, num_filters*2, (3, 3))
    # the feature map to size of (batch, grid_h, grid_w, num_anchors*(classes+5))
    y = darknetconv2d(y, out_filters, (1, 1), num_classes=num_classes)
    # Reshape the output size (batch, grid_h, grid_w, num_anchors, classes+5)
    y = YoloOutputLayer(num_anchors, num_classes)(y)
    return x, y
```

- The setting of the convolutional layer of YOLO-v3

```
                                              Code path
                                          model→darknet.py
layer_count = 1
def darknetconv2d(x, filters, kernel_size, strides=(1, 1), num_classes=80):
    global layer_count
    padding = 'valid' if strides == (2, 2) else 'same'
```

```
x = layers.Conv2D(filters, kernel_size, strides,
        padding=padding,
        kernel_regularizer=tf.keras.regularizers.l2(5e-4),
        name='conv2d_last_layer{}_{}'.format(layer_count, num_classes))(x)
layer_count += 1
    return x
```

- YoloOutputLayer: This custom layer is used to reshape the output size to the size of (bacth, grid_h, grid_w, anchors, classes+5)

```
class YoloOutputLayer(tf.keras.layers.Layer):
    def __init__(self, num_anchors, num_classes, **kwargs):
        super(YoloOutputLayer, self).__init__(**kwargs)
        self.num_anchors = num_anchors
        self.num_classes = num_classes

    def build(self, input_shape):
        self.input_h, self.input_w = input_shape[1:3]

    def call(self, x, **kwargs):
        if self.input_h is None or self.input_w is None:
            x = tf.reshape(x, (-1, tf.shape(x)[1], tf.shape(x)[2],
                    self.num_anchors, self.num_classes + 5))
        else:
            x = tf.reshape(x, (-1, self.input_h, self.input_w,
                    self.num_anchors, self.num_classes + 5))
        return x
```

```
Code path
layers→output_layer.py
```

- YoloOutputBoxLayer: convert the predicted coordinates (t_x, t_y, t_w, t_h) of the model into bounding box (x, y, w, h) through the following formula:

$$b_x = \sigma(t_x) + c_x$$
$$b_y = \sigma(t_y) + c_y$$
$$b_w = p_w e^{t_w}$$
$$b_h = p_h e^{t_h}$$

b_x, b_y, b_w, b_h: coordinates of the center, width, and height of the predicted bounding box, respectively
c_x, c_y: top left corner of the image
σ: sigmoid activation function
p_w, p_h: width and height of the anchor box, respectively

```python
class YoloOutputBoxLayer(tf.keras.layers.Layer):
    def __init__(self, anchors, output_layer=1,
            num_classes=80, training=False, **kwargs):
        super(YoloOutputBoxLayer, self).__init__(**kwargs)
        self.anchors = anchors
        self.num_classes = num_classes
        self.training = training
        # The output size is (13, 13), multiply by 32 to restore the original input size (416,
        416)
        if output_layer == 1:
            self.grid_to_img_scale = 32
        elif output_layer == 2:
            self.grid_to_img_scale = 16
        else:
            self.grid_to_img_scale = 8


    def build(self, input_shape):
        self.grid_h, self.grid_w = input_shape[1:3]


    def call(self, inputs, **kwargs):
        """
        :param inputs:  (batch, grid_h, grid_w, anchors, [x, y, w, h, obj, ...classes])
        :param kwargs: None
        :return:
            bbox: (batch, grid_h, grid_w, anchors, [x1, y1, x2, y2])
            box_confidence: (batch, grid_h, grid_w, anchors, 1)
            box_class_probs: (batch, grid_h, grid_w, anchors, classes)
        """
        # use tf.shape to dynamically obtain the output size of the previous layer
        if self.grid_h is None:
            grid_h, grid_w = tf.shape(inputs)[1], tf.shape(inputs)[2]
        else:
            grid_h, grid_w = self.grid_h, self.grid_w


        box_xy, box_wh, box_confidence, box_class_probs = \
            tf.split(inputs, (2, 2, 1, self.num_classes), axis=-1)
```

Code path
layers→output_box_layer.py

```python
# box_xy: (batch, grid_h, grid_w, anchors, [tx, ty])
box_xy = tf.sigmoid(box_xy)     # scale to 0~1
# box_confidence: (batch, grid_h, grid_w, anchors, confidence)
box_confidence = tf.sigmoid(box_confidence)     # scale to 0~1
# box_class_probs: (batch, grid_h, grid_w, anchors, classes)
box_class_probs = tf.sigmoid(box_class_probs)  # scale to 0~1
# pred_box: (batch, grid_h, grid_w, anchors, [tx,ty,tw,th])
pred_box = tf.concat((box_xy, box_wh), axis=-1)


grid = tf.meshgrid(tf.range(grid_w), tf.range(grid_h))
grid = tf.stack(grid, axis=-1)        # (gx, gy, 2)
grid = tf.expand_dims(grid, axis=2)     # (gx, gy, 1, 2)


# b_x = σ(t_x) + c_x, b_y = σ(t_y) + c_y
box_xy = (box_xy + tf.cast(grid, tf.float32)) / tf.cast((grid_w, grid_h), tf.float32)
# Calculate input image size
img_w, img_h=(grid_w*self.grid_to_img_scale, grid_h*self.grid_to_img_scale)
# b_w = p_w e^{t_w}, b_h = p_h e^{t_h}
box_wh = self.anchors * tf.exp(box_wh) / (img_w, img_h)


# bbox: (x1, y1, x2, y2)
box_x1y1 = box_xy - box_wh / 2
box_x2y2 = box_xy + box_wh / 2
bbox = tf.concat([box_x1y1, box_x2y2], axis=-1)
if self.training:
    return tf.concat([bbox,box_confidence,box_class_probs,pred_box],axis=-1)
return bbox, box_confidence, box_class_probs
```

- NMSLayer: This custom layer is used to remove the redundant bounding box prediction

```python
class NMSLayer(tf.keras.layers.Layer):
    """

    Non maximum suppression Layer
    """

    def __init__(self, num_classes, iou_threshold, score_threshold, **kwargs):
        super(NMSLayer, self).__init__(**kwargs)
        self.num_classes = num_classes
        self.iou_threshold = iou_threshold
        self.score_threshold = score_threshold
```

Code path
layers→nms_layer.py

```python
def call(self, inputs, **kwargs):
    """
    :param inputs: [OutputLayer1, OutputLayer2, OutputLayer3]
    :return:
        boxes: (batch, 100, 4)
        scores: (batch, 100)
        classes: (batch, 100)
        valid_detections: (batch)
    """
    bboxes, box_conf, box_class = [], [], []
    # Bboxes of the three output layers
    for pred in inputs:
        bboxes.append(tf.reshape(pred[0], (tf.shape(pred[0])[0], -1, 4)))
        box_conf.append(tf.reshape(pred[1], (tf.shape(pred[1])[0], -1, 1)))
        box_class.append(tf.reshape(pred[2],
                    (tf.shape(pred[2])[0], -1, self.num_classes)))
    bboxes = tf.concat(bboxes, axis=1)
    box_conf = tf.concat(box_conf, axis=1)
    box_class = tf.concat(box_class, axis=1)


    # Prediction box score
    scores = box_conf * box_class
    # Remove redundant boxes
    boxes, scores, classes, valid_detections = \
        tf.image.combined_non_max_suppression(
            boxes=tf.reshape(bboxes, (tf.shape(bboxes)[0], -1, 1, 4)),
            scores=tf.reshape(scores, (tf.shape(scores)[0], -1, self.num_classes)),
            max_output_size_per_class=100,
            max_total_size=100,
            iou_threshold=self.iou_threshold,
            score_threshold=self.score_threshold)
    return boxes, scores, classes, valid_detections
```

- YOLO-v3 loss function

    The source code of the objective loss function is described as below:

```python
def yolo_loss(y_true, y_pred, anchors, num_classes=80,
        ignore_thresh=0.5):
    """
    :param y_true: (batch_size, grid_h, grid_w, anchors, [x1, y1, x2, y2, obj, ...cls])
    :param y_pred: (batch_size, grid_h, grid_w, anchors, [x, y, w, h, obj, ...cls])
    :param anchors: three anchors box shape: (3, 2)
    :param num_classes: number of classes in dataset
    :param ignore_thresh: if (IoU < threshold) and ignore
    :return: total loss (xy_loss + wh_loss + confidence_loss + class_loss)
    """
```

Code path
losses→yolo_loss.py

```
"""
# 1. Convert prediction output
# y_pred: (batch, grid_h, grid_w, anchors, [x1, y1, x2, y2, obj, ...classes, tx, ty, tw, th])
    # y_pred is the output of the YoloOutputBoxLayer layer, which is divided into 4 parts
    here:
    # pred_box: (batch, grid_h, grid_w, anchors, [x1, y1, x2,y2]) is used to calculate the
    IoU with the ground truth box
# pred_obj: (batch, grid_h, grid_w, anchors, obj) is used to calculate confidence_loss
# pred_class: (batch, grid_h, grid_w, anchors, classes) is used to calculate class_loss
    # pred_xywh: (batch, grid_h, grid_w, anchors, [tx, ty, tw, th]) is used to calculate
    xy_loss, wh_loss
pred_box, pred_obj, pred_class, pred_xywh = tf.split(y_pred, (4, 1, num_classes, 4),
                                                    axis=-1)
pred_xy = pred_xywh[..., 0:2]
pred_wh = pred_xywh[..., 2:4]


# 2. Convert (x1, y1, x2, y2)▨  (x, y, w, h)
true_box, true_obj, true_class_idx = tf.split(y_true, (4, 1, 1), axis=-1)
true_xy = (true_box[..., 0:2] + true_box[..., 2:4]) / 2
true_wh = true_box[..., 2:4] - true_box[..., 0:2]
    # Because the calculated loss value of small objects is small, the small objects are
    multiplied by a larger "weight factor"
    box_loss_scale = 2 - true_wh[..., 0] * true_wh[..., 1]


# 3. Convert (x1, y1, x2, y2) to (tx, ty, tw, th) and calculate the loss with pred_xywh
grid_h, grid_w = tf.shape(y_true)[1], tf.shape(y_true)[2]
grid = tf.meshgrid(tf.range(grid_w), tf.range(grid_h))
grid = tf.expand_dims(tf.stack(grid, axis=-1), axis=2)
true_xy = true_xy * (grid_h, grid_w) - tf.cast(grid, true_xy.dtype)
true_wh = tf.math.log(true_wh / anchors)
true_wh = tf.where(tf.math.is_inf(true_wh), tf.zeros_like(true_wh), true_wh)


# 4. Generate mask (with or without objects), shape(batch_size, grid, grid, anchors)
obj_mask = tf.squeeze(true_obj, -1)


# 5. Generate negative sample mask, ignore false positive if iou exceeds threshold
# Get the bounding box in which the object exists, true_box_flat = (N, [x1, y1, x2, y2])
true_box_flat = tf.boolean_mask(true_box, tf.cast(obj_mask, tf.bool))
# Calculate iou of ground-truth box and predicted box
best_iou = tf.reduce_max(broadcast_iou(pred_box, true_box_flat), axis=-1)
```

```
# Generate mask, if iou <ignore_thresh
ignore_mask = tf.cast(best_iou < ignore_thresh, tf.float32)


# 5. Calculate the loss function
xy_loss = obj_mask * box_loss_scale * tf.reduce_sum(tf.square(true_xy - pred_xy),
                                                    axis=-1)

wh_loss = obj_mask * box_loss_scale * tf.reduce_sum(tf.square(true_wh - pred_wh),
                                                    axis=-1)
obj_loss = binary_crossentropy(true_obj, pred_obj)
confidence_loss = obj_mask * obj_loss + (1 - obj_mask) * ignore_mask * obj_loss
class_loss = obj_mask * sparse_categorical_crossentropy(true_class_idx, pred_class)


    # 6. Sum the loss values of all prediction boxes (batch, grid, grid, anchors) => (batch,
    1)
xy_loss = tf.reduce_sum(xy_loss, axis=(1, 2, 3))
wh_loss = tf.reduce_sum(wh_loss, axis=(1, 2, 3))
confidence_loss = tf.reduce_sum(confidence_loss, axis=(1, 2, 3))
class_loss = tf.reduce_sum(class_loss, axis=(1, 2, 3))
return xy_loss + wh_loss + confidence_loss + class_loss


# Wrap the yolo loss function by the" tf.keras.losses.LossFunctionWrapper"
class YoloLoss(tf.keras.losses.LossFunctionWrapper):
    def __init__(self,
            anchors,
            num_classes=80,
            ignore_thresh=0.5,
            name='yolo_loss'):
        super(YoloLoss, self).__init__(
            yolo_loss,
            name=name,
            anchors=anchors,
            num_classes=num_classes,
            ignore_thresh=ignore_thresh)
```

The function for calculation of IoU

```
def broadcast_iou(pred_box, true_box):
    """
    Calculate the IoU between the ground-truth box and the predicted box
    :param pred_box: size(b, gx, gy, 3, 4)
```

Code path
losses→yolo_loss.py

```
:param true_box: size(n, 4)
:return: Intersection over Union(IoU)
"""
# broadcast boxes
pred_box = tf.expand_dims(pred_box, -2)   # (b, gx, gy, 3, 1, 4)
true_box = tf.expand_dims(true_box, 0)    # (1, n, 4)
# new_shape: (b, gx, gy, 3, n, 4)
new_shape = tf.broadcast_dynamic_shape(tf.shape(pred_box), tf.shape(true_box))
pred_box = tf.broadcast_to(pred_box, new_shape)
true_box = tf.broadcast_to(true_box, new_shape)


# Overlap: (b, gx, gy, 3, n)
int_w = tf.maximum(tf.minimum(pred_box[..., 2], true_box[..., 2]) -
            tf.maximum(pred_box[..., 0], true_box[..., 0]), 0)
int_h = tf.maximum(tf.minimum(pred_box[..., 3], true_box[..., 3]) -
            tf.maximum(pred_box[..., 1], true_box[..., 1]), 0)
int_area = int_w * int_h


# box size: w * h
box_1_area = (pred_box[..., 2] - pred_box[..., 0]) * (pred_box[..., 3] - pred_box[..., 1])
box_2_area = (true_box[..., 2] - true_box[..., 0]) * (true_box[..., 3] - true_box[..., 1])
return int_area / (box_1_area + box_2_area - int_area)
```

- Data preprocessing

The source code for data preprocessing is stored at "utils→dataset.py." Data preprocessing for training the YOLO-v3 model is divided into two steps: (1) data augmentation and (2) training target transformation.

- Data augmentation: color conversion, horizontal flip, image scale, and image rotation

```
def parse_aug_fn(dataset, input_size=(416, 416)):        Code path
    """                                                  utils→dataset.py
    Data Augmentation
    """

    ih, iw = input_size
    # 1) Prepare information
    # x shape: (None, None, 3)
    x = tf.cast(dataset['image'], tf.float32) / 255.  # Normalization
    # bbox shape: (y1, x1, y2, x2)
```

```
bbox = dataset['objects']['bbox']
# label shape: (1,)
label = tf.cast(dataset['objects']['label'], tf.float32)
# Adjust the image to a fixed size (at the same time adjust the bounding box)
x, bbox = resize(x, bbox, input_size)


# 2) Data augmentation
    # color conversion
    x = tf.cond(tf.random.uniform([], 0, 1) > 0.75, lambda: color(x), lambda: x)
# flipping image
x, bbox = tf.cond(tf.random.uniform([], 0, 1) > 0.5,
                  lambda: flip(x, bbox), lambda: (x, bbox))
# image scale
x, bbox, label = tf.cond(tf.random.uniform([], 0, 1) > 0.5,
                         lambda: zoom(x, bbox, label), lambda: (x, bbox, label))
# image rotation
x, bbox, label = tf.cond(tf.random.uniform([], 0, 1) > 0.5,
                         lambda: rotate(x, bbox, label), lambda: (x, bbox, label))


# 3) Data Integration
# shape: (num_boxes, [x1, y1, x2, y2, classes])
y = tf.stack([bbox[1], bbox[0], bbox[3], bbox[2], label], axis=-1)
# Normalize (x1, y1, x2, y2)
y = tf.divide(y, [ih, iw, ih, iw, 1])
# extend (num_boxes, [x1, y1, x2, y2, classes]) to (100, [x1, y1, x2, y2, classes])
paddings = [[0, 100 - tf.shape(y)[0]], [0, 0]]
y = tf.pad(y, paddings)
# Redefine the shape of output y
y = tf.ensure_shape(y, (100, 5))
return x, y
```

- Training target transformation: Transform data format (batch size, 100, [x1, y1, x2, y2, classes]) into (batch size, grid, grid, anchors, [x, y, w, h, obj, class])

```
def transform_targets(x_train, y_train, anchors,                    Code path
                      anchor_masks, grid_size=13):               utils→dataset.py
    """
    transform y_label to training target label,
    (batch, 100, [x1, y1, x2, y2, class])→(batch, grid, grid, anchor, [x, y, w, h, obj, class])
```

```
:param x_train: shape: (None, 416, 416, 3)
:param y_train: shape: (None, 100, [x1, y1, x2, y2, class])
:param anchors: 9 preset anchors boxes, shape: (9,2)
:param anchor_masks: the mask of the anchors box
:return:
    x_train: training image, shape: (batch, img_h, img_w, 3)
    y_outs: Return the training data output by three different layers
        ((batch, grid, grid, 3, [x, y, w, h, obj, class, best_anchor]),
         (batch, grid, grid, 3, [x, y, w, h, obj, class, best_anchor]),
         (batch, grid, grid, 3, [x, y, w, h, obj, class, best_anchor]))
"""

y_outs = []
# calculate anchor index for true boxes
anchors = tf.cast(anchors, tf.float32)
# The area of the anchor boxes
anchor_area = anchors[..., 0] * anchors[..., 1]
# Calculate the area of ground-truth boxes
box_wh = y_train[..., 2:4] - y_train[..., 0:2]
box_wh = tf.tile(tf.expand_dims(box_wh, -2),
        (1, 1, tf.shape(anchors)[0], 1))
box_area = box_wh[..., 0] * box_wh[..., 1]
# Calculate the IoU of ground truth boxes objects and anchor boxes
intersection = tf.minimum(box_wh[..., 0], anchors[..., 0]) * \
    tf.minimum(box_wh[..., 1], anchors[..., 1])
iou = intersection / (box_area + anchor_area - intersection)
# Get the largest anchor box index value
anchor_idx = tf.cast(tf.argmax(iou, axis=-1), tf.float32)
anchor_idx = tf.expand_dims(anchor_idx, axis=-1)


# y_train: (batch, 100, [x1, y1, x2, y2, classes])
# anchor_idx: (batch, 100, best_anchor)
# combine y_train and anchor_idx-> (batch, 100, [x1, y1, x2, y2, classes, best_anchor])
y_train = tf.concat([y_train, anchor_idx], axis=-1)


# generate output training data for three different layers
for anchor_idxs in anchor_masks:
    y_outs.append(transform_targets_for_output(y_train, grid_size, anchor_idxs))
    grid_size *= 2
return x_train, tuple(y_outs)
```

```python
@tf.function
def transform_targets_for_output(y_true, grid_size, anchor_idxs):
    """
    Generate a training label for the output layer
    (batch, 100, [x1, y1, x2, y2, class, best_anchor])->
    (batch, grid, grid, anchor, [x, y, w, h, obj, class])
    :param y_true: shape: (N, boxes, (x1, y1, x2, y2, class, best_anchor))
    :param grid_size: grid cell size
    :param anchor_idxs: The index value of the anchor boxes of each layer (each layer has
    3 anchor boxes)
    :return:(batch, grid, grid, anchor, [x, y, w, h, obj, class])
    """
    batch = tf.shape(y_true)[0]
    y_true_out = tf.zeros((batch, grid_size, grid_size, tf.shape(anchor_idxs)[0], 6))
    anchor_idxs = tf.cast(anchor_idxs, tf.int32)
    indexes = tf.TensorArray(tf.int32, 1, dynamic_size=True)
    updates = tf.TensorArray(tf.float32, 1, dynamic_size=True)
    idx = 0
    for i in tf.range(batch):
        for j in tf.range(tf.shape(y_true)[1]):
            if tf.equal(y_true[i][j][2], 0):
                continue
            anchor_eq = tf.equal(anchor_idxs, tf.cast(y_true[i][j][5], tf.int32))
            if tf.reduce_any(anchor_eq):
                # box: (x1, y1, x2, y2)
                box = y_true[i][j][0:4]
                box_xy = (y_true[i][j][0:2] + y_true[i][j][2:4]) / 2
                anchor_idx = tf.cast(tf.where(anchor_eq), tf.int32)
                grid_xy = tf.cast(box_xy // (1/grid_size), tf.int32)
                indexes = indexes.write(idx,
                        [i, grid_xy[1], grid_xy[0], anchor_idx[0][0]])
                updates = updates.write(idx,
                        [box[0], box[1], box[2], box[3], 1, y_true[i][j][4]])
                idx += 1
    return tf.tensor_scatter_nd_update(y_true_out, indexes.stack(), updates.stack())
```

- Converting pre-trained model

  Using "convert.py" file to convert the original pre-trained YOLO-v3 model into the HDF5 format used by Keras.

Step 1: Download the trained model weights of the original YOLO-v3:

```
wget https://pjreddie.com/media/files/yolov3.weights -O model_data/yolov3.weights
```

Step 2: Convert the trained model by using "convert.py" file, a "yolov3.h5" file is generated.

```
python convert.py
```

- Setting hyperparameters for training

    The "config.py" file is used to store hyperparameters for training the YOLO-v3 model.

```
import numpy as np                                          Code path
                                                            config.py

# Yolo Anchor boxes size
yolo_anchors = np.array([(10, 13), (16, 30), (33, 23), (30, 61), (62, 45),
                (59, 119), (116, 90), (156, 198), (373, 326)], np.float32)
# Yolo Anchor boxes mask, ex: 6,7,8 for the first output layer, 3,4,5 for the second output
layer
yolo_anchor_masks = np.array([[6, 7, 8], [3, 4, 5], [0, 1, 2]])
# Input size of the YOLO
size_h = 416
size_w = 416


# Training is divided into two steps
step1_batch_size = 32
step1_learning_rate = 1e-3
step1_start_epochs = 0
step1_end_epochs = 100
step2_batch_size = 8
step2_learning_rate = 1e-4
step2_start_epochs = step1_end_epochs
step2_end_epochs = step1_end_epochs + 100


# Pre-Trained weights
yolo_weights = 'model_data/yolo_weights.h5'


# Classes of COCO dataset (the order of classes following yolo_weights.h5')
coco_classes = ['person', 'bicycle', 'car', 'motorcycle', 'airplane', 'bus', 'train', 'truck', 'boat',
```

'traffic light', 'fire hydrant', 'stop sign', 'parking meter', 'bench', 'bird', 'cat',

'dog', 'horse', 'sheep', 'cow', 'elephant', 'bear', 'zebra', 'giraffe', 'backpack',

'umbrella', 'handbag', 'tie', 'suitcase', 'frisbee', 'skis', 'snowboard', 'sports ball',

'kite', 'baseball bat', 'baseball glove', 'skateboard', 'surfboard', 'tennis racket',

'bottle', 'wine glass', 'cup', 'fork', 'knife', 'spoon', 'bowl', 'banana', 'apple',

'sandwich', 'orange', 'broccoli', 'carrot', 'hot dog', 'pizza', 'donut', 'cake',

'chair', 'couch', 'potted plant', 'bed', 'dining table', 'toilet', 'tv', 'laptop', 'mouse',

'remote', 'keyboard', 'cell phone', 'microwave', 'oven', 'toaster', 'sink',

'refrigerator', 'book', 'clock', 'vase', 'scissors', 'teddy bear', 'hair drier',

'toothbrush']

 # the classes of VOC dataset
voc_classes = ['aeroplane', 'bicycle', 'bird', 'boat', 'bottle', 'bus', 'car', 'cat', 'chair', 'cow',

'diningtable', 'dog', 'horse', 'motorbike', 'person', 'pottedplant', 'sheep', 'sofa',

'train', 'tvmonitor']

(b) Building and training the YOLO-v3 model

To train the YOLO-v3 model, the converted YOLO-v3 model weights (yolov3.h5) are loaded first, and then the model is fine-tuned on the PASCAL VOC 2007 dataset. The code is written in the "train.py" file.

▪ Import packages

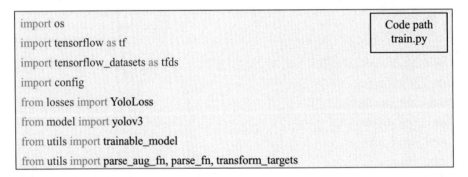

```
import os
import tensorflow as tf
import tensorflow_datasets as tfds
import config
from losses import YoloLoss
from model import yolov3
from utils import trainable_model
from utils import parse_aug_fn, parse_fn, transform_targets
```
Code path
train.py

▪ Set hyperparameter for training YOLO-v3

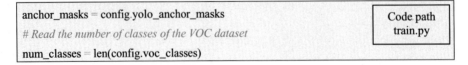

```
anchor_masks = config.yolo_anchor_masks
# Read the number of classes of the VOC dataset
num_classes = len(config.voc_classes)
```
Code path
train.py

▪ Create YOLO-v3 model and load weights

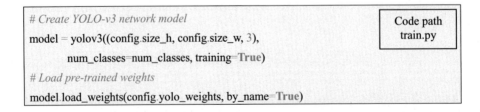

```
# Create YOLO-v3 network model
model = yolov3((config.size_h, config.size_w, 3),
        num_classes=num_classes, training=True)
# Load pre-trained weights
model.load_weights(config.yolo_weights, by_name=True)
```
Code path
train.py

- Set callback

```
# Create logs directory
log_dir = 'logs_yolo'
model_dir = log_dir + '/models'
os.makedirs(model_dir, exist_ok=True)
# Save training log
model_tb = tf.keras.callbacks.TensorBoard(log_dir=log_dir)
# Save the best model weights
model_mckp=tf.keras.callbacks.ModelCheckpoint(model_dir+'/best_{epoch:03d}.h5',
                        monitor='val_loss',
                        save_best_only=True,
                        mode='min')
# After 10 epochs, if the val_loss does not improve, learning rate is reduced
mdoel_rlr = tf.keras.callbacks.ReduceLROnPlateau(verbose=1)
```

Code path
train.py

- Creating function for training

```
def training_model(model, callbacks, num_classes=80, step=1):
    # Set training parameters
    if step == 1:
        batch_size = config.step1_batch_size
        learning_rate = config.step1_learning_rate
        start_epochs = config.step1_start_epochs
        end_epochs = config.step1_end_epochs

    else:
        batch_size = config.step2_batch_size
        learning_rate = config.step2_learning_rate
        start_epochs = config.step2_start_epochs
        end_epochs = config.step2_end_epochs
    anchors = config.yolo_anchors / 416

    AUTOTUNE = tf.data.experimental.AUTOTUNE # Auto adjustment mode
```

Code path
train.py

```python
# combine data
combined_split = tfds.Split.TRAIN + tfds.Split.VALIDATION
train_data, info = tfds.load("voc2007", split=combined_split, with_info=True)
# shuffle data
train_data = train_data.shuffle(1000)
# Data standardization and data augmentation,
train_data = train_data.map(lambda dataset: parse_aug_fn(dataset),
                            num_parallel_calls=AUTOTUNE)
# batch size
train_data = train_data.batch(batch_size)
# Training target conversion
train_data = train_data.map(lambda x,y:transform_targets(x,y,anchors,anchor_masks),
            num_parallel_calls=AUTOTUNE)
# Enable prefetch mode
train_data = train_data.prefetch(buffer_size=AUTOTUNE)


# Validation data
# Use test set as verification data, a total of 4952 images
val_data = tfds.load("voc2007", split=tfds.Split.TEST)
val_data = val_data.map(lambda dataset: parse_fn(dataset),
            num_parallel_calls=AUTOTUNE)
# batch size
val_data = val_data.batch(batch_size)
# target transformation
val_data = val_data.map(lambda x, y: transform_targets(x, y, anchors, anchor_masks),
            num_parallel_calls=AUTOTUNE)
# Enable prefetch mode
val_data = val_data.prefetch(buffer_size=AUTOTUNE)


# Set the optimizer
optimizer = tf.keras.optimizers.Adam(lr=learning_rate)
# Set loss function
model.compile(optimizer=optimizer,
        loss=[YoloLoss(anchors[mask],
                                num_classes=num_classes) for mask in anchor_masks],
        run_eagerly=False)
# Train the network model
model.fit(train_data,
    epochs=end_epochs,
    callbacks=callbacks,
```

```python
    validation_data=val_data,
    initial_epoch=start_epochs)
```

- Training YOLO-v3 (Step 1): Freeze all the layers of the model; only the weights of the output layers to be updated

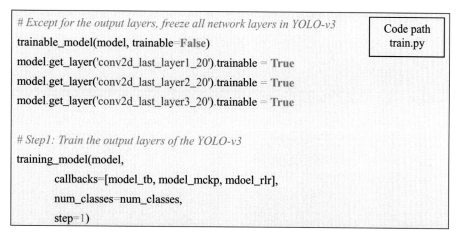

```
# Except for the output layers, freeze all network layers in YOLO-v3
trainable_model(model, trainable=False)
model.get_layer('conv2d_last_layer1_20').trainable = True
model.get_layer('conv2d_last_layer2_20').trainable = True
model.get_layer('conv2d_last_layer3_20').trainable = True


# Step1: Train the output layers of the YOLO-v3
training_model(model,
        callbacks=[model_tb, model_mckp, mdoel_rlr],
        num_classes=num_classes,
        step=1)
```

Code path
train.py

- Training YOLO-v3 (Step 2): Train entire network

```
# Unfreeze all network layers of YOLO-v3
trainable_model(darknet, trainable=True)
# Train the entire YOLO-v3 network model
print("Start teraining Step2")
training_model(model,
        callbacks=[model_tb, model_mckp, mdoel_rlr, model_ep],
        num_classes=num_classes,
        step=2)
```

Code path
train.py

(c) Making predictions

After training, the trained YOLO-v3 weights on the Pascal VOC 2007 training set are used for making predictions on the VOC 2007 test set. The code is written in the "test.py" file.

- Import packages

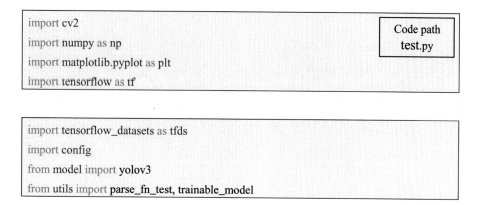

```
import cv2
import numpy as np
import matplotlib.pyplot as plt
import tensorflow as tf
```

Code path
test.py

```
import tensorflow_datasets as tfds
import config
from model import yolov3
from utils import parse_fn_test, trainable_model
```

- Load test data and data information

```
# Use a total of 4952 data in the test set
test_data = tfds.load("voc2007", split=tfds.Split.TEST)
weight_file = 'model_data/yolo_weights.h5'  # or 'logs_yolo/models/best_xxx.h5'


if weight_file == 'model_data/yolo_weights.h5':
    # COCO weights
    classes_list = config.coco_classes
    num_classes = len(config.coco_classes)
    freeze = False
else:
    # VOC2007 weights
    classes_list = config.voc_classes
    num_classes = len(config.voc_classes)
    if int(os.path.splitext(weight_file)[0].split('_')[-1]) <= 100:
        freeze = True
```

Code path
test.py

- Create YOLO-v3 model and load weights

```
# Create YOLO-v3 model
model = yolov3((config.size_h,config.size_w,3),
            num_classes=num_classes training=False)
# Specify the model weights , please modify the file name "best_xxx.h5"
weight_file = 'logs_yolo/models/best_xxx.h5'


# If the model weights is in step1, freeze all the layers Darknet-53
if int(os.path.splitext(weight_file)[0].split('_')[-1]) <= 100:
    darknet = model.get_layer('Yolo_DarkNet')
    trainable_model(darknet, trainable=False)


# load model weights
model.load_weights(weight_file)
```

Code path
test.py

- Detection of YOLO-v3

```
def test_and_show_result(model, test_number=10):
    for data in test_data.take(test_number):
        # Reading data
        org_img = data['image'].numpy()
        h, w, _ = data['image'].shape
        # Pre-processing the data
        img, bboxes = parse_fn_test(data)
        # Predicted boxes and score
        boxes, scores, classes, nums = model.predict(tf.expand_dims(img, axis=0))
        boxes, scores, classes, nums = boxes[0], scores[0], classes[0], int(nums[0])
        for i in range(nums):
            # Mark the predicted bounding boxes on the image
            x1y1 = tuple((np.array(boxes[i][0:2]) * (w, h)).astype(np.int32))
            x2y2 = tuple((np.array(boxes[i][2:4]) * (w, h)).astype(np.int32))
            cv2.rectangle(org_img, x1y1, x2y2, (255, 0, 0), 2)
            # Show the predicted object category on the image
            cv2.putText(org_img,
                '{} {:.4f}'.format(classes_list[int(classes[i])], scores[i]),
                x1y1,
                cv2.FONT_HERSHEY_SIMPLEX,
                1, (255, 0, 0), 2)
    plt.figure()
    plt.imshow(org_img)
plt.show()
```

Code path
test.py

- Detection results: Run the function for making prediction; Fig. 12.39 shows some detection results of the YOLO-v3 model.

```
test_and_show_result(model, test_number=1)
```

FIG. 12.39 Detection results of the YOLO-v3 model on the Pascal VOC test set.

References

[1] X. Zhu, Y. Wang, J. Dai, L. Yuan, Y. Wei, Flow-guided feature aggregation for video object detection, in: 2017 IEEE International Conference on Computer Vision, Venice, 2017, pp. 408–417.

[2] S.C. Huang, An advanced motion detection algorithm with video quality analysis for video surveillance systems, IEEE Trans. Circuits Syst. Video Technol. 21 (1) (2011) 1–14.

[3] S.C. Huang, B.H. Do, Radial basis function based neural network for motion detection in dynamic scenes, IEEE Trans. Cybern. 44 (1) (2014) 114–125.

[4] Y. Taigman, M. Yang, M. Ranzato, L. Wolf, Deepface: closing the gap to human-level performance in face verification, Proc. IEEE Conf. Comput. Vis. Pattern Recognit. (2014) 1701–1708.

[5] F. Schroff, D. Kalenichenko, J. Philbin, Facenet: a unified embedding for face recognition and clustering, Proc. IEEE Conf. Comput. Vis. Pattern Recognit. (2015) 815–823.

[6] S.C. Huang, M.K. Jiau, C.A. Hsu, A high-efficiency and high-accuracy fully automatic collaborative face annotation system for distributed online social networks, IEEE Trans. Circuits Syst. Video Technol. 24 (10) (2014) 1810–1813.

[7] L. Zheng, Y. Yang, Q. Tian, SIFT meets CNN: a decade survey of instance retrieval, IEEE Trans. Pattern Anal. Mach. Intell. 40 (5) (2018) 1224–1244.

[8] A. Gordo, J. Almazan, J. Revaud, D. Larlus, Deep image retrieval: learning global representations for image search, in: European Conference on Computer Vision, 2016, pp. 241–257.

[9] K. Cao, A.K. Jain, Automated latent fingerprint recognition, IEEE Trans. Pattern Anal. Mach. Intell. 41 (4) (2019) 788–800.

[10] A. Gangwar, A. Joshi, DeepIrisNet: deep iris representation with applications in iris recognition and cross-sensor iris recognition, in: 2016 IEEE International Conference on Image Processing, Phoenix, AZ, 2016, pp. 2301–2305.

[11] S. Huang, B. Chen, Highly accurate moving object detection in variable bit rate video-based traffic monitoring systems, IEEE Trans. Neural Netw. Learn. Syst. 24 (12) (2013) 1920–1931.

[12] S. Huang, B. Chen, Automatic moving object extraction through a real-world variable-bandwidth network for traffic monitoring systems, IEEE Trans. Ind. Electron. 61 (4) (2014) 2099–2112.

[13] Y. Liu, D. Jaw, S. Huang, J. Hwang, DesnowNet: context-aware deep network for snow removal, IEEE Trans. Image Process. 27 (6) (2018) 3064–3073.

[14] B. Chen, S. Huang, C. Li, S. Kuo, Haze removal using radial basis function networks for visibility restoration applications, IEEE Trans. Neural Netw. Learn. Syst. 29 (8) (2018) 3828–3838.

[15] S. Huang, B. Chen, W. Wang, Visibility restoration of single hazy images captured in real-world weather conditions, IEEE Trans. Circuits Syst. Video Technol. 24 (10) (2014) 1814–1824.

[16] T.-H. Le, P.-H. Lin, S.-C. Huang, LD-Net: an efficient lightweight denoising model based on convolutional neural network, IEEE Open, J. Comput. Soc. 1 (2020) 173–181.

[17] S. Huang, J. Ye, B. Chen, An advanced single-image visibility restoration algorithm for real-world hazy scenes, IEEE Trans. Ind. Electron. 62 (5) (2015) 2962–2972.

[18] B. Chen, S. Huang, Edge collapse-based Dehazing algorithm for visibility restoration in real scenes, J. Disp. Technol. 12 (9) (2016) 964–970.

[19] B. Chen, S. Huang, S. Kuo, Error-optimized sparse representation for single image rain removal, IEEE Trans. Ind. Electron. 64 (8) (2017) 6573–6581.

[20] R. Girshick, J. Donahue, T. Darrell, J. Malik, Rich feature hierarchies for accurate object detection and semantic segmentation, Proc. IEEE Conf. Comput. Vis. Pattern Recognit. (2014) 580–587.

[21] J. Uijlings, K. van de Sande, T. Gevers, A. Smeulders, Selective search for object recognition, Int. J. Comput. Vis. (2013).

[22] R. Girshick, Fast R-CNN, in: Proceedings of the IEEE International Conference on Computer Vision, 2015, pp. 1440–1448.

[23] S. Ren, K. He, R.B. Girshick, J. Sun, Faster R-CNN: towards real-time object detection with region proposal networks, in: Advances in Neural Information Processing Systems, 2015, pp. 91–99.

[24] J. Redmon, S. Divvala, R. Girshick, A. Farhadi, You only look once: unified, real-time object detection, Proc. IEEE Conf. Comput. Vis. Pattern Recognit. (2016) 779–788.

[25] W. Liu, D. Anguelov, D. Erhan, C. Szegedy, S.E. Reed, Ssd: single shot multibox detector, in: European Conference on Computer Vision, 2015, pp. 21–37.

[26] J. Redmon, A. Farhadi, YOLO9000: better, faster, stronger, Proc. IEEE Conf. Comput. Vis. Pattern Recognit. (2017) 7263–7271.

[27] T.-Y. Lin, P. Dollár, R. Girshick, K. He, B. Hariharan, S. Belongie, Feature pyramid networks for object detection, Proc. IEEE Conf. Comput. Vis. Pattern Recognit. (2017) 2117–2125.

[28] T.-Y. Lin, P. Goyal, R. Girshick, K. He, P. Dollar, Focal loss for dense object detection, in: Proceedings of the IEEE International Conference on Computer Vision, 2017, pp. 2999–3007.

[29] K. He, G. Gkioxari, P. Dollar, R.B. Girshick, Mask R-CNN, in: Proceedings of the IEEE International Conference on Computer Vision, 2017, pp. 2980–2988.

[30] J. Redmon, A. Farhadi, YOLOv3: An incremental improvement, arXiv preprint arXiv:1804.02767, (2018).

[31] T. Le, S. Huang, D. Jaw, Cross-resolution feature fusion for fast hand detection in intelligent homecare systems, IEEE Sensors J. 19 (12) (2019) 4696–4704.

[32] S. Huang, T. Le and D. Jaw, "DSNet: joint semantic learning for object detection in inclement weather conditions," in IEEE Transactions on Pattern Analysis and Machine Intelligence, https://doi.org/10.1109/TPAMI.2020.2977911.

[33] T.-H. Le, D.-W. Jaw, I.-C. Lin, H.-B. Liu, S.-C. Huang, An efficient hand detection method based on convolutional neural network, in: 2018 7th International Symposium on Next Generation Electronics (ISNE), IEEE, 2018, pp. 1–2.

[34] C. Szegedy, et al., Going deeper with convolutions, in: IEEE Conference on Computer Vision and Pattern Recognition, 2015, pp. 1–9.

[35] K. He, X. Zhang, S. Ren, J. Sun, Deep residual learning for image recognition, Proc. IEEE Conf. Comput. Vis. Pattern Recognit. (2016) 770–778.

[36] J. Long, E. Shelhamer, T. Darrell, Fully convolutional networks for semantic segmentation, Proc. IEEE Conf. Comput. Vis. Pattern Recognit. (2015) 3431–3440.

[37] L.-C. Chen, G. Papandreou, F. Schroff, H. Adam, Rethinking atrous convolution for semantic image segmentation, arXiv preprint arXiv:1706.05587, (2017).

[38] Y. Li, H. Qi, J. Dai, X. Ji, Y. Wei, Fully convolutional instance-aware semantic segmentation, Proc. IEEE Conf. Comput. Vis. Pattern Recognit. (2017) 2359–2367.

[39] Z.-Q. Zhao, P. Zheng, S.-t. Xu, X. Wu, Object detection with deep learning: a review, IEEE Trans. Neural Netw. Learn. Syst. 30 (11) (2019) 3212–3232.

[40] J.R. Uijlings, K.E. Van De Sande, T. Gevers, A.W. Smeulders, Selective search for object recognition, Int. J. Comput. Vis. 104 (2) (2013) 154–171.

[41] S. Fidler, R. Mottaghi, A. Yuille, R. Urtasun, Bottom-up segmentation for top-down detection, in: Proceedings of the IEEE Conference on Computer Vision and Pattern Recognition, 2013, pp. 3294–3301.

[42] P. Sermanet, D. Eigen, X. Zhang, M. Mathieu, R. Fergus, Y. LeCun, Overfeat: Integrated recognition, localization and detection using convolutional networks, arXiv preprint arXiv:1312.6229, (2013).

[43] M.D. Zeiler, R. Fergus, Visualizing and understanding convolutional networks, in: European Conference on Computer Vision, 2014, pp. 818–833.

[44] K. Simonyan, A. Zisserman, Very deep convolutional networks for large-scale image recognition, in: International Conference on Learning Representations, 2015, pp. 1–14.

[45] L. Chen, et al., DeepLab: Semantic image segmentation with deep convolutional nets, atrous convolution, and fully connected CRFs, IEEE Trans. Pattern Anal. Mach. Intell. 40 (4) (2018) 834–848.

Index

Note: Page numbers followed by *f* indicate figures *t* indicate tables and *b* indicate boxes.

Printed in the United States
by Baker & Taylor Publisher Services